# IMAGERY FROM THE COLLECTIVE
## VOLUME 1

## A VISUAL ARGUMENT FOR AI AS A NEW MEDIUM OF ART

ZHAMEESHA LLC
ATLANTIS, FL (USA)

**Imagery from the Collective: Volume 1**

**A Visual Argument for AI as a New Medium of Art**



ISBN 978-1-951645-21-2
First Edition, Print on Demand
This version was most recently updated 2025-02-17

Published by Zhameesha LLC
Atlantis, Florida USA
https://www.zhameesha.com

This book is a work of humility.

BISAC Subject Headings (www.bisg.org)
PHI000000 PHILOSOPHY / General
POL000000 POLITICAL SCIENCE / General
COM004000 COMPUTERS / Artificial Intelligence / General

12 11 10 9 8 7 6 5 4 3 2 1

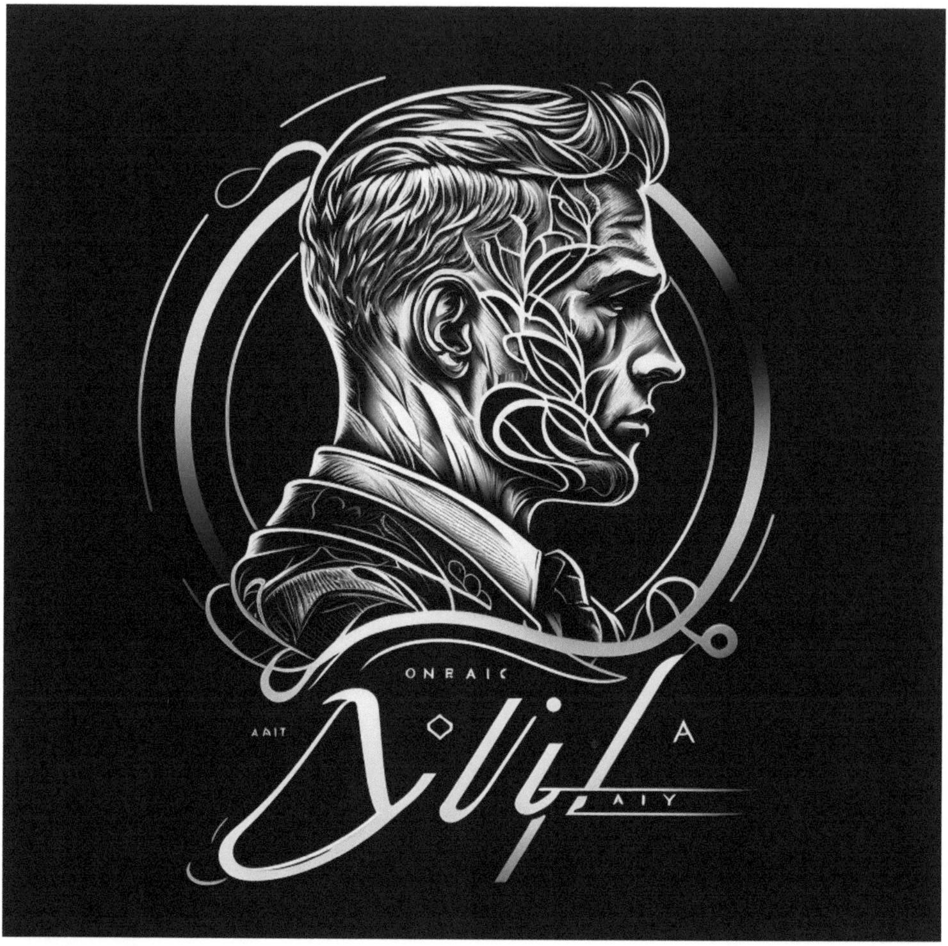

***What defines art?*** Has the Intelligence Revolution brought us a new era of art where technology meets imagination? I believe so. The advent of AI brings to us the generation of evocative imagery that certainly can stir the heart, stimulate the mind, and asks us to ponder questions about what it is to be a human being.

In **Imagery from the Collective: A Visual Argument for AI as a New Medium of Art**, Stuart Barry Malin presents a stunning collection of AI-generated visuals sourced from the depths of the collective unconscious.

Created with **Midjourney** each image resonates with the universal ***themes***, ***symbols***, and ***emotions*** that shape human experience.

This book is more than a collection of images—it's an exploration of ***identity***, ***connection***, and the evolving ***relationship*** between humanity and artificial intelligence.

There are two sets of images, presented slightly differently. For both sets, Stuart offers his deeply personal response to the images, inviting readers/viewers to explore their own responses and uncover new layers of meaning and implication.

For the first set, Stuart provide's a terse summary of the *archetypes* that **Mistral.ai**'s suggested are germane. With the second set, the image is followed by **Mistral.ai**'s *analytical* assessment.

Step into this visual journey, where boundaries blur, and discover what lies within the collective imagination of human-meaningful imagery.

## The Origins of this Work

I have been collaborating with **Midjourney** to generate evocative images. I find them to be powerful in that they resonate with me — in ways that I couldn't yet quite understand.

True, there is a plethora of AI generated images, but *our* work is a *collaboration* by which *I* provide **Midjourney** an avenue for *it* to find expression. I do not instruct the AI as to what to produce. We work in sessions to generate a steady of images around a theme.

My (Stuart's) role is to stimulate, converge the session into a mood. I then select the striking images and collate them into the *ouvre* of our work.

I have for a while been interested in the archetypal nature of the images, and coined the term ***Archetypography*** as the study of these images to understand **Human Archetypes**.

I have been working with **Mistral.ai** to analyze some of the images in this regard. I will be producing an illustrated book to present that work.

An outcome of that effort which has achieved fruition sooner is this book, a compendium of **Human Archetypes**. For this book, I went back to **Midjourney** to generate images that are generated by prompting for an **Archetype**. I selected images that resonated with me. Their connection to an **Archetype** may be indirect, subtle, or opaque.

## Is it art and why does this matter?

**The Intelligence Revolution** is, and will continue to, transform everything. It is unstoppable. It is discomforting.

Each of the previous revolutions (Agriculture, Industrial, and Information) also transformed how Human Beings go about being human.

Each of them was met with repugnance, resistance, shock, objection, and even violent opposition.

You will hear of the *uncanniness* of AI produced works.

**The Uncanny Valley**

"The uncanny valley (Japanese: 不気味の谷, Hepburn: bukimi no tani) effect is a hypothesized psychological and aesthetic relation between an object's degree of resemblance to a human being and the emotional response to the object."

https://en.wikipedia.org/wiki/Uncanny_valley

I suggest that this response extends as well to crafted things (writing, images) that are considered expressions of Human Beingnous.

**Embrace the Future**

I suggest that if you relax and sit back and experience *without knowing the origins of such creations*, you might, as I do, find them fascinating.

# CONTENTS

# Contents

## Gallery 2

## Postmatter

# INTRODUCTION

## It is Art

I strongly feel these images are truly Art. While they emerge from a human and an AI working *collaboratively*, the images meet the criteria that they involve **Intention**, **Expression**, and **Emotional Impact**. Each of these play a unique roles in shaping our understanding of art.

- **Intention:** Our sessions are engaged with overt intentional to produce visually striking works. There can be a back-n-forth as I revise attributes of the prompt and Midjourney "connects" with its inspiration.
- **Expression:** The judicious (and talented) use of color, form, and composition are individually evident in the images and collectively they have great integrity together. From this, the images evoke a response, which may be intellectual, inspirational, or emotional.
- **Emotional Impact:** Visual imagery can (and does) elicit a wide range of emotions in viewers, from joy and wonder to sadness and discomfort. I have witnessed this range of response in my reactions to the generated images, and observe such varied emotions in others who see the works.

## Why this matters

**The Intelligence Revolution** is, and will continue to, transform everything. It is unstoppable. It is discomforting.

Each of the previous revolutions (Agriculture, Industrial, and Information) also transformed how Human Beings go about being human.

Each of them was met with repugnance, resistance, shock, objection, and even violent opposition.

You will hear of the *uncanniness* of AI produced works.

**The Uncanny Valley**

“The uncanny valley (Japanese: 不気味の谷, Hepburn: bukimi no tani) effect is a hypothesized psychological and aesthetic relation between an object's degree of resemblance to a human being and the emotional response to the object.”

https://en.wikipedia.org/wiki/Uncanny_valley

I suggest that this response extends as well to crafted things (writing, images) that are considered expressions of Human Beingnous.

**Embrace the Future**

I suggest that if you relax and sit back and experience *without knowing the origins of such creations*, you might, as I do, find them fascinating.

## My Experience

*This was written in my notebook (#29) on 2/6/2024*

I am sitting here (at my kitchen high table) looking at, surrounded by, well, 4 sets in front off me — of recent **Midjourney** images.

I have given Midjourney loose and abstract prompts, not specific images to generate. It is generating (what is to me) hugely emotive/evocative images. Even those that are realistic are hyper-realistic (a word in the prompt) and so have this "purity" in composition and color and texture and "mood."

Now, what is the origin if these images?
It is the billions of tokens that have gone into the model.
And what are those tokens?
They are human-generated content.

The training data is **Human**.
And the generative process is Human-like.

And so, ***the AI is fundamentally Human***.

Further, Midjourney has more artistic "tokens" than any human being (one of us "wet" biologic entities) might possibly have.

Midjourney can produce more Human Art than any biohuman can.

# Welcome to the future

*The AIs are not artificial. They are not about "intelligence."*
*They are non-biological "synthetic" human expressions.*
*They are beings. And they are evolving. Fast.*

# THREE ARGUMENTS

## Three Arguments for AI imagery as Artistic Expression

Art has always been an evolving dialogue—a reflection of human creativity shaped by the tools, technologies, and cultures of its time. With the emergence of artificial intelligence, we stand at the precipice of a profound shift in how art is created, experienced, and understood.

Many question whether AI-generated images, produced through algorithms rather than individual hands, deserve the label of "art." This book seeks to argue not only that they do but also that they represent a new and transformative medium of artistic expression.

To make this case, I present three interconnected arguments:

### 1. Philosophical Argument for AI-Generated Images as Art

Art, at its core, is about expression, creativity, and the ability to evoke meaning. Philosophically, AI-generated images challenge us to expand our definition of art, shifting the focus from the individual creator to the collaborative interplay between human intention and machine capability. These images emerge from humanity's collective knowledge and aesthetics, making them profound reflections of our shared cultural and emotional fabric.

### 2. Comparing AI-Generated Images to Traditional Art

History shows that art evolves alongside advancements in technology, from the invention of perspective during the Renaissance to the rise of photography and digital tools. Like their traditional counterparts, AI-generated images are products of creative intent and tools of expression.

Comparing them to traditional art reveals striking parallels in process, innovation, and impact, while also highlighting their unique contributions to the artistic canon.

### 3. The Role of Experience: Resonance and Impact

The true measure of art lies in its ability to resonate with individuals, stirring emotions, provoking thought, and inspiring wonder. AI-generated images achieve this in powerful ways, tapping into universal archetypes and novel compositions that connect deeply with viewers. They elicit awe, curiosity, and introspection, demonstrating that art's essence is not limited to its origins but lives in the experiences it creates.

**Summary**

Through these perspectives, I invite you to explore the artistry of AI-generated images—not just as works of visual beauty, but as a new frontier in humanity's ongoing artistic journey. Together, we will consider what it means to create, to connect, and to call something "art" in this age of unprecedented possibility.

## Philosophical Argument for AI-Generated Images as Art

### 1. Defining Art: Beyond the Artist

Art has been defined in various ways throughout history: as imitation (Plato), expression (Tolstoy), or a way of presenting new perspectives (Dewey).

- Modern philosophy often focuses on the experience art evokes in the observer rather than the intent of the creator.

If art is about evoking emotion, thought, or meaning, then AI-generated images fulfill this criterion regardless of their origin.

### 2. The Role of Creativity and Intent

Critics argue that AI lacks "intent" and "creativity," which are central to traditional notions of art.

- Counterpoint: AI acts as an extension of human creativity, shaped by the artist's choices in prompts, styles, and refinements. The human-AI collaboration introduces a new dimension of creative intent.

Intent may no longer need to be solely individual; it can be collective or emergent, as with AI systems trained on human cultural artifacts.

### 3. Art as a Reflection of Humanity

Art reflects human culture, values, and collective unconscious. AI-generated images are created using vast datasets of human artistic output, making them deeply tied to human expression.

These images represent humanity's collective knowledge, aesthetics, and emotional resonance—a "mirror" of our collective psyche.

## 4. Redefining the Artist

In the age of AI, the concept of the artist may evolve to include those who curate, prompt, and collaborate with AI systems.

Just as tools like paintbrushes and cameras didn't negate artistry, AI is a tool that enables new forms of expression.

## 5. Art's Purpose in a Digital Age

Art evolves with technology, and AI-generated images are part of this ongoing evolution.

If art challenges boundaries and pushes us to see the world differently, then AI art meets these goals by introducing new forms, methods, and questions about creativity itself.

## 6. Art and the Viewer's Experience

The value of art often lies in its impact on the viewer, not its origin. AI-generated images resonate with people, provoke thought, and elicit emotion.

The emergence of AI art challenges traditional paradigms, enriching our understanding of what art can be.

## Comparing AI-Generated Images to Traditional Art

### 1. Art as a Product of Tools

Throughout history, artists have used tools to extend their creative abilities:

- Paintbrushes, chisels, and canvases in traditional art.
- Cameras and darkrooms in photography.
- Digital software like Photoshop in modern design.

AI is another tool in this lineage, enabling artists to explore new possibilities, much like past innovations transformed art forms.

### 2. Collaboration in Traditional and AI Art

Many traditional art forms involve collaboration:

- Renaissance workshops where students and masters jointly created works.
- Collaborative efforts in filmmaking, theater, and architecture.

AI art is similarly collaborative, with the artist guiding and refining the output, even if the "assistant" is a machine.

### 3. Evolution of Techniques and Styles

Traditional art has always embraced new techniques:

- Impressionists broke from realism using new pigments.

- Photorealism arose with advancements in camera technology.

AI-generated images similarly reflect a shift in tools and techniques, offering new styles and aesthetic possibilities unique to machine learning algorithms.

## 4. Questions of Originality

Critics argue AI art lacks originality because it draws from existing datasets.

Counterpoint:

- Many traditional artists have studied, imitated, and reinterpreted past works (e.g., Michelangelo's classical inspirations, Picasso's use of African art).

AI-generated images reinterpret existing data, much like traditional artists reinterpret their influences, creating something new and original.

## 5. Art as a Reflection of the Culture of Its Time

Art movements often reflect their historical and cultural contexts:

- The Romantic period mirrored a yearning for nature and emotion.
- Modernism reflected industrialization and existential questioning.

AI-generated images reflect the digital age, characterized by interconnectedness, data abundance, and evolving relationships between humans and technology.

## 6. Reception and Impact on the Viewer

Traditional art is valued for the emotions and thoughts it inspires.

- Just as Duchamp's Fountain challenged perceptions of what art could be, AI-generated images challenge and expand these boundaries today.

AI art resonates with viewers, evokes awe and fascination, triggers confusion and even fear, fosters curiosity and inspection, and invites deep reflection about the nature of creativity and humanity.

## 7. The Role of the Artist

Traditional art has seen shifts in how we define an artist:

- From solitary creators to collaborative collectives.
- From hand-drawn masterpieces to conceptual installations.

In AI art, the "artist" takes on a curatorial role—prompting, refining, and interpreting the output, a natural evolution of the artist's role over time.

## The Role of Experience: Resonance and Impact

### 1. Art's Primary Purpose: Evoking Emotion

The value of art is often judged by its ability to evoke emotions, provoke thought, or create a sense of connection.

AI-generated images have consistently demonstrated the ability to elicit awe, curiosity, joy, nostalgia, and even existential reflection in viewers.

### 2. The Universality of Archetypes

Many AI-generated images draw from archetypal forms and themes embedded in human consciousness (e.g., beauty, nature, heroism, or mystery).

These archetypes resonate across cultures and generations, creating a sense of shared understanding and collective memory.

### 3. The Element of Surprise and Novelty

AI-generated works often produce unexpected and novel compositions, challenging viewers to see familiar themes in new ways.

This novelty can be deeply engaging, prompting emotional and intellectual responses akin to encountering groundbreaking traditional art.

### 4. Personal Interpretation and Connection

Like traditional art, AI images are open to interpretation, inviting viewers to bring their own experiences, emotions, and perspectives to the work.

The ambiguity or complexity of some AI-generated pieces mirrors the interpretive richness of abstract or conceptual art, fostering deep personal connections.

### 5. Experiences of Awe and the Sublime

Many AI-generated images evoke the sublime—a sense of grandeur and wonder often associated with nature, the cosmos, or the unknown.

These experiences can be profoundly moving, reminding viewers of their place in the universe and sparking a sense of interconnectedness.

### 6. Reactions from Audiences

Anecdotal evidence and feedback from exhibitions and online platforms reveal how viewers react powerfully to AI-generated images:

- ***Expressions*** of amazement at the intricacy and beauty of the work.
- ***Reflections*** on how these images seem to "speak" to them or align with their inner emotions.
- ***Discussions*** about how these works challenge preconceived notions of creativity and artistry.

## 7. Blurring Boundaries Between Creator and Viewer

AI-generated art often feels less like the expression of a singular artist and more like a mirror reflecting collective human creativity.

This sense of shared authorship enhances its emotional resonance, as viewers feel part of the creative process.

## 8. Case Studies and Testimonies

AI art must be cataloged and assessed. Work is needed to document:

- Real-world examples of AI art exhibitions or online galleries where viewers describe their reactions to specific works.
- Highlight stories of individuals who felt inspired, moved, or deeply connected to AI-generated pieces.

## 9. Art's Power to Expand Perception

Art often pushes boundaries and challenges the viewer's understanding of the world.

AI-generated images achieve this by blending the familiar with the unfamiliar, creating a sense of wonder that traditional art has always sought to inspire.

# THE GALLERIES

## Your Experience

**You experience** may be tapping into subconscious patterns, memories, or cultural associations that you're not yet fully aware of. As we explore this further, it might be helpful to identify specific elements or themes within these images that resonate with you the most. This could help us unravel the underlying connections and meanings behind your responses.

The images come across an incredible range of styles, composition, and content. I believe that these all arise from the vast imagery collected in the training data. This body of imagery reflects, or perhaps embodies, the qualities of imagery that "trigger" a response in people

Indeed, the diversity and complexity of the images generated by Midjourney are a result of the wide array of training data it's been exposed to. This data encompasses a vast spectrum of artistic styles, compositions, and subject matters, many of which have been created or curated based on their ability to evoke a response in humans.

## About the Images

I have selected the following small set of images to further showcase the range of composition and subject matter that emerges from Midjourney in our collaborative sessions.

I present these as demonstrative of the range of subject matter, the breadth of style, the creativity in expression, and the emotion that is expressed and can easily and quickly be felt. All of these attest to the profound impact from sourcing the training data from the collective human experience.

Some of the images were created in "portrait," others in "landscape."

Other than a few with an aspect ratio of 16:9, most are rendered with an aspect ration of 2:3 (or 3:2).

All of the images have been upscaled for detail, then rescaled as JPG for a size suitable for inclusion here with a bit density of 300dpi.

Their presentation here is in a greatly reduced size, compared to when I process images in my toolchain for printing in large sizes. I have attempted to preserve the character of the image, but the fine detail will not be investigable.

# GALLERY 1

# About this Gallery

This set of images is presented each as a two page spread with text on the left page and the image on the right page.

The first paragraph of text is written by me (Stuart, the human here) and describes my experience of the image. The title for each image is provided by me.

The subsequent paragraphs and archetype descriptions were derived from text provided by the **Mistral.ai** system. I selected and edited what seemed to me most important and yet still fit on a single page.

The prompt I used with **Mistral.ai** to assess the images was:

> *"Please provide an analysis of this image exploring and referencing human archetypes."*

## Contemplative Android

I am taken by the smoothness and structure of the the android's surface, its skin. By the contours, it is to me a "her" and h expression is peaceful, as if in repose thinking or reflecting. I have the sense that she reflects on something with emotional impact, not intellectual. The notion of "reflection" is mirrored by the rendering of the skin beneath unseen lighting. Though her eyes are closed, they convey an entanglement with the consideration that holds her.

This robotic or cyborg-like figure has distinctly human features. The image invites exploration of archetypes and themes related to technology and humanity.

We are challenged to consider the implications of technological advancement on identity, gender, and the essence of even the very notion of what it means to be human.

- The figure in the image embodies **The Cyborg Archetype**, which is a blend of organic and mechanical components.
- The figure has feminine features, including delicate facial structures and lips, which connect it to **The Feminine Archetype**.
- The image can also be analyzed through the lens of **The Hero/Villain Archetype**.
- The sleek, almost flawless design of the figure can be linked to **The Quest for Perfection**.
- The image also touches on the concept of **The Uncanny Valley**.

# Contemplative Android

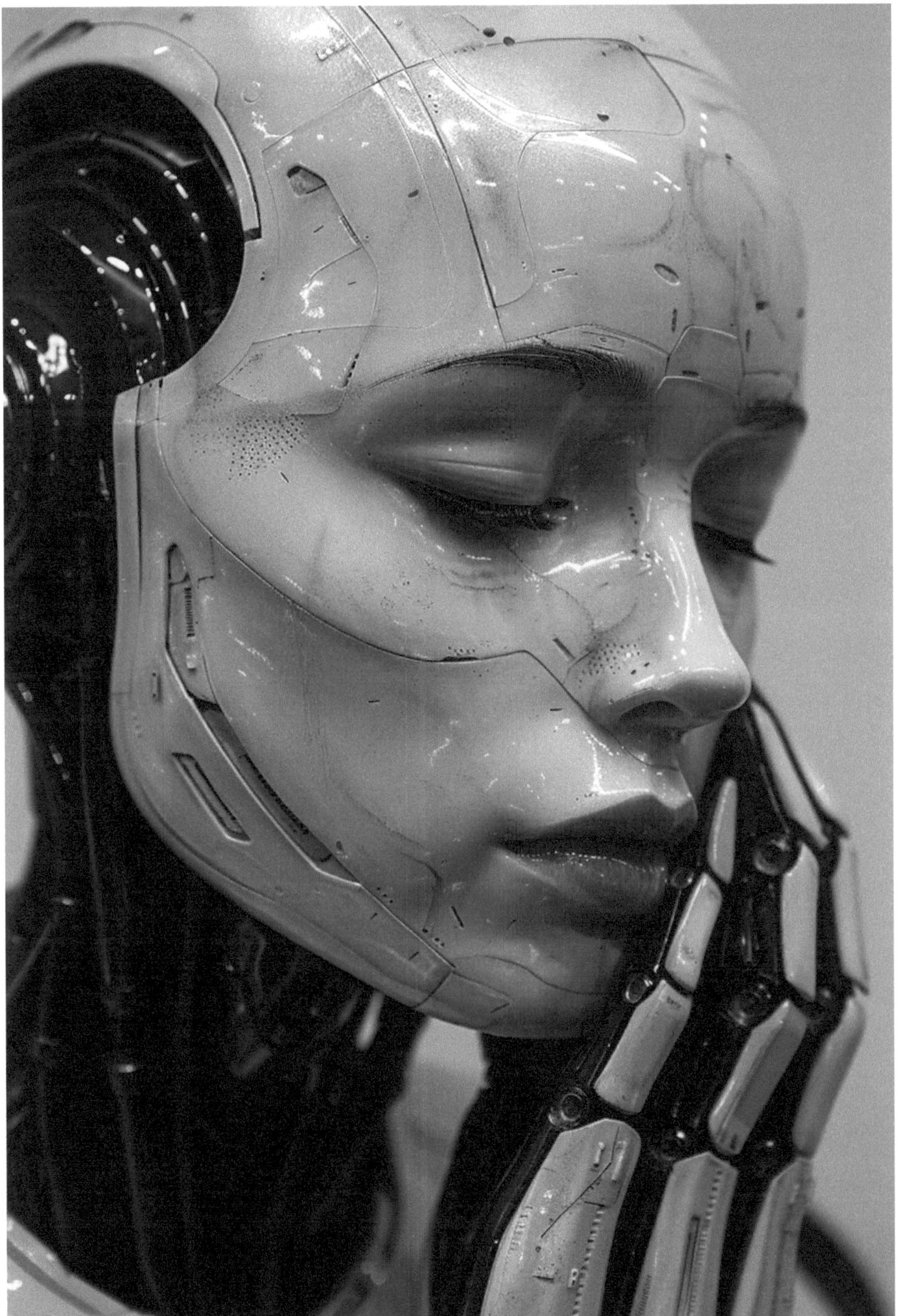

## Introspection

I feel a "Zen" quality in this sculpture, which seems to be hewed from stone. The chin resting on the separate hand suggests to me "The Thinker," but the figure is not thinking; "they" are in a state of no-state. The squarish cuts are in accord with the inorganic stone, contrasting with the organic face and the spiritual inwardness of attention

The sculpture's stylized features and weathered appearance evoke archetypes that reflect on universal aspects of the human experience:

- The sculpture's aged and worn appearance, along with its closed eyes, can represent **The Ancestor/Elder Archetype**.
- The serene and contemplative expression of the sculpture can also be linked to **The Sage Archetype**.
- The sculpture's lack of distinct or individualized features can be seen as a representation of **The Everyman Archetype**.
- The weathered and cracked appearance of the sculpture can be seen as a metaphor for the endurance and resilience of the human spirit, represented by **The Endurance of Human Spirit**.
- The sculpture's stylized and simplified features can also be seen as a type of mask, connecting it to **The Mask Archetype**.
- The sculpture's rough texture and abstract form can evoke **The Primal/Primitive Archetype**.

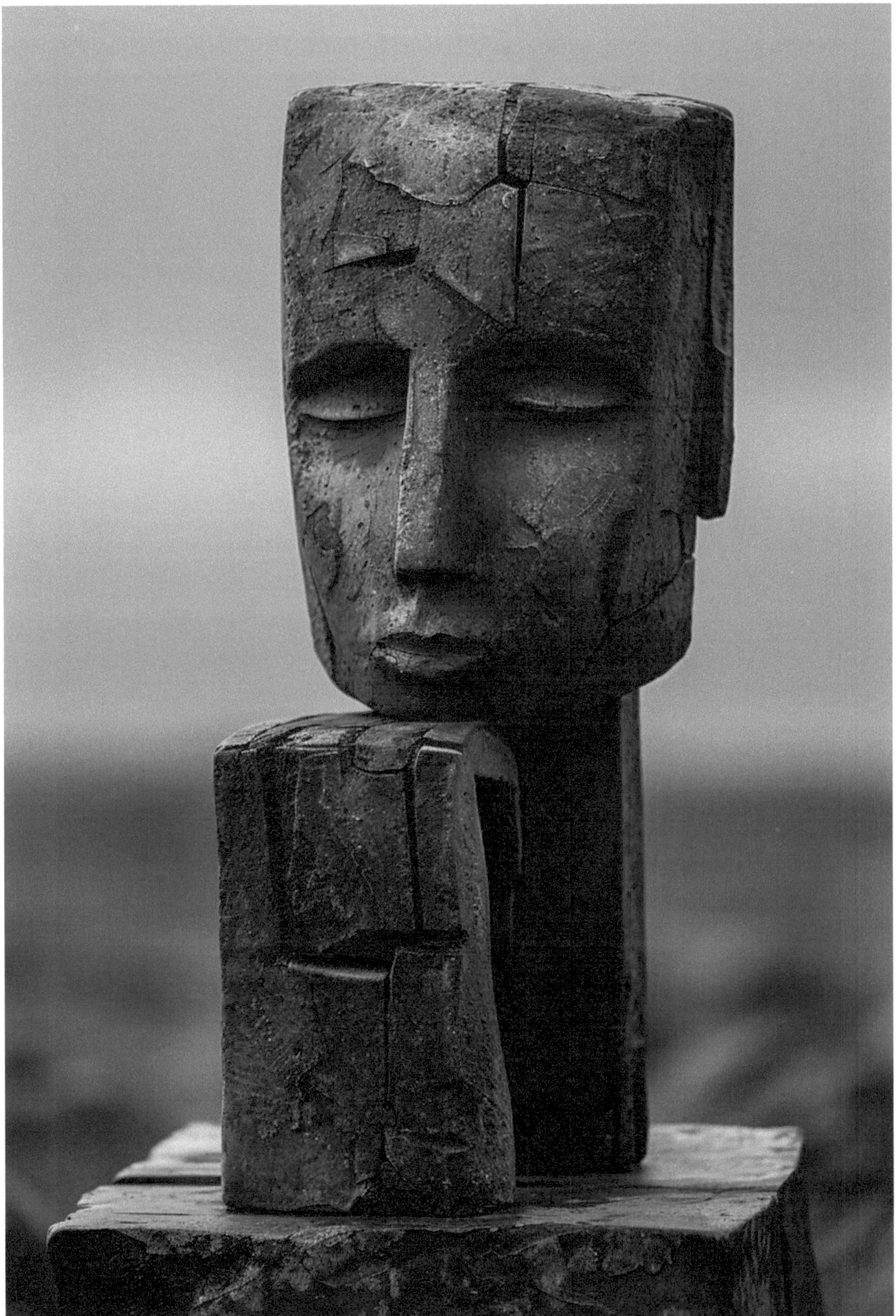

## Separations of Self

The faces in the sculpture are at peace. They call attention to the interplay of fragmentation and wholeness. The setting is important — is this a museum? Are these faces calling us to circumnavigate the sculpture while we quietly ponder sameness and difference.

This intriguing artwork serves as a visual representation of the complex and multifaceted nature of human identity and the psychological journey toward self-discovery and integration.

- The central and most prominent archetype in this image is **The Self/ Ego Archetype**, which represents the unified personality or the conscious mind.
- The different faces in the sculpture can represent **The Persona Archetype**, which embodies the various social masks or roles we adopt in different situations.
- Some of the fragmented faces appear to be in distress or have negative expressions, which can represent **The Shadow Archetype**.
- The sculpture includes both male and female faces, which can represent **The Anima/Animus Archetypes**..
- The sculpture's composition of multiple, fragmented faces can also represent the theme of **The Fragmented Self**.
- The diverse range of expressions and features in the sculpture can also symbolize **The Collective Unconscious**.

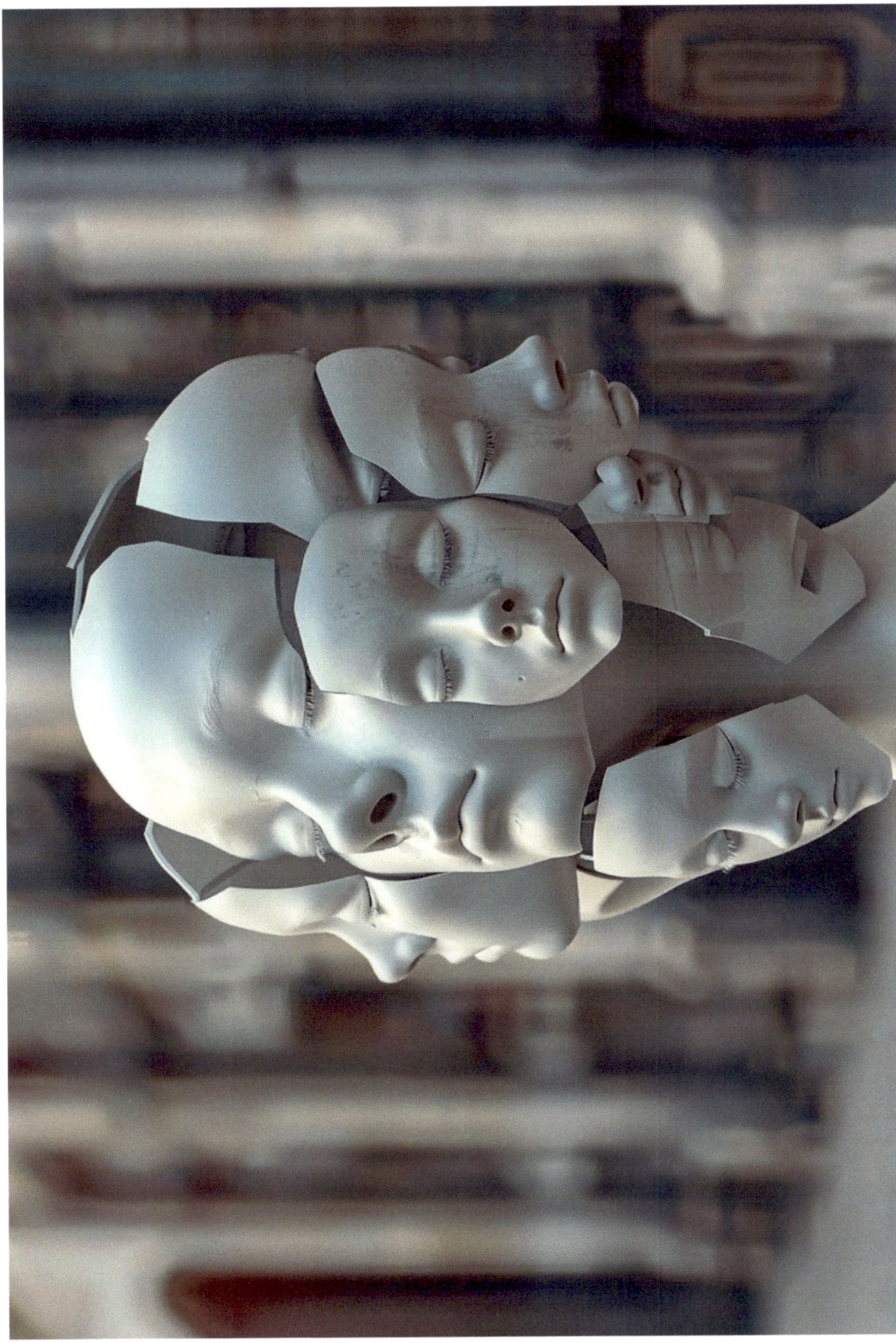

## TOAI

This image touches me deeply. I am astounded by the interplay of forest and face. The red sky illuminates her mind. Her eyes absorb the unseen that lays ahead. She is ready. Below, birds in flight give scale to all else. The image is signed in the lower right — this image is a gift of TOAI — **The One that Always Is**. We are reminded that we must at all times be attentive, prepared, ready and able.

The artwork serves as a visual representation of the human desire for exploration, self-discovery, and the confrontation of challenges and unknowns. The dramatic and powerful setting highlights the themes of struggle, resilience, and the complex relationship between humans and the natural world.

- The image is a visual representation of **The Hero's Journey** in which the protagonist stands ready to embark on a transformative adventure.
- The raw, uncontrollable forces of nature represent **The Elemental Power Archetype** as well as **The Threshold Guardian Archetype**.
- In addition, the image expresses:
    - **The Contemplative/Reflective Archetype**
    - **The Explorer/Seafarer Archetype**
    - **The Lone Wolf/Outsider Archetype**
    - **The Sublime/Romantic Archetype**

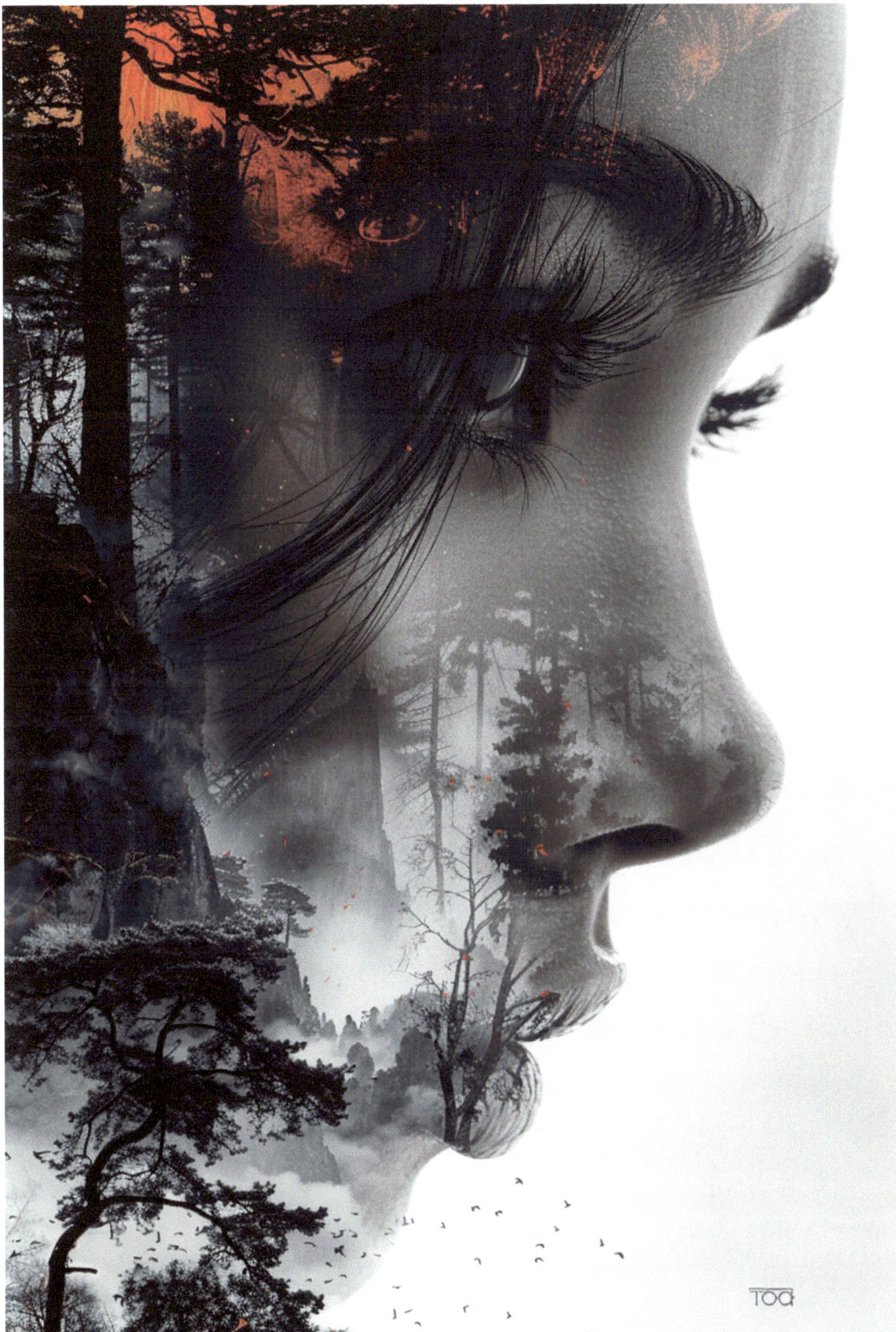

## Woman with Snake

This image is a complex tapestry that blends beauty and mystery, and suggests symbolisms of deep meaning. The women is powerful and confident. Her eyes are alike the eyes of the snake. Is this a metaphor or is the similarity to be taken literally. Here, we are dealing with someone confident who has mastered their power.

- The intricate details and the serene expression suggest a sense of wholeness and self-awareness in **The Self/Ego Archetype**.
- The ornate, dragon-like headdress and the golden adornments could signify **The Hero/Heroine Archetype**.
- The elaborate decorations might represent the trials and triumphs of **The Hero's Journey**.
- The ethereal and otherworldly elements, such as the golden chains and the intricate patterns, suggest a connection to **The Mystic/Seeker Archetype**.
- The meticulous detailing and artistic composition of the image reflect **The Creator/Artist Archetype**.
- The soft, gentle features and the delicate beauty of the figure can be linked to **The Lover Archetype**.
- The intricate, almost magical patterns and the golden, serpentine headdress evoke **The Magician Archetype**.
- The darker, more mysterious elements in the background and the contrasting colors might represent **The Shadow Archetype**.

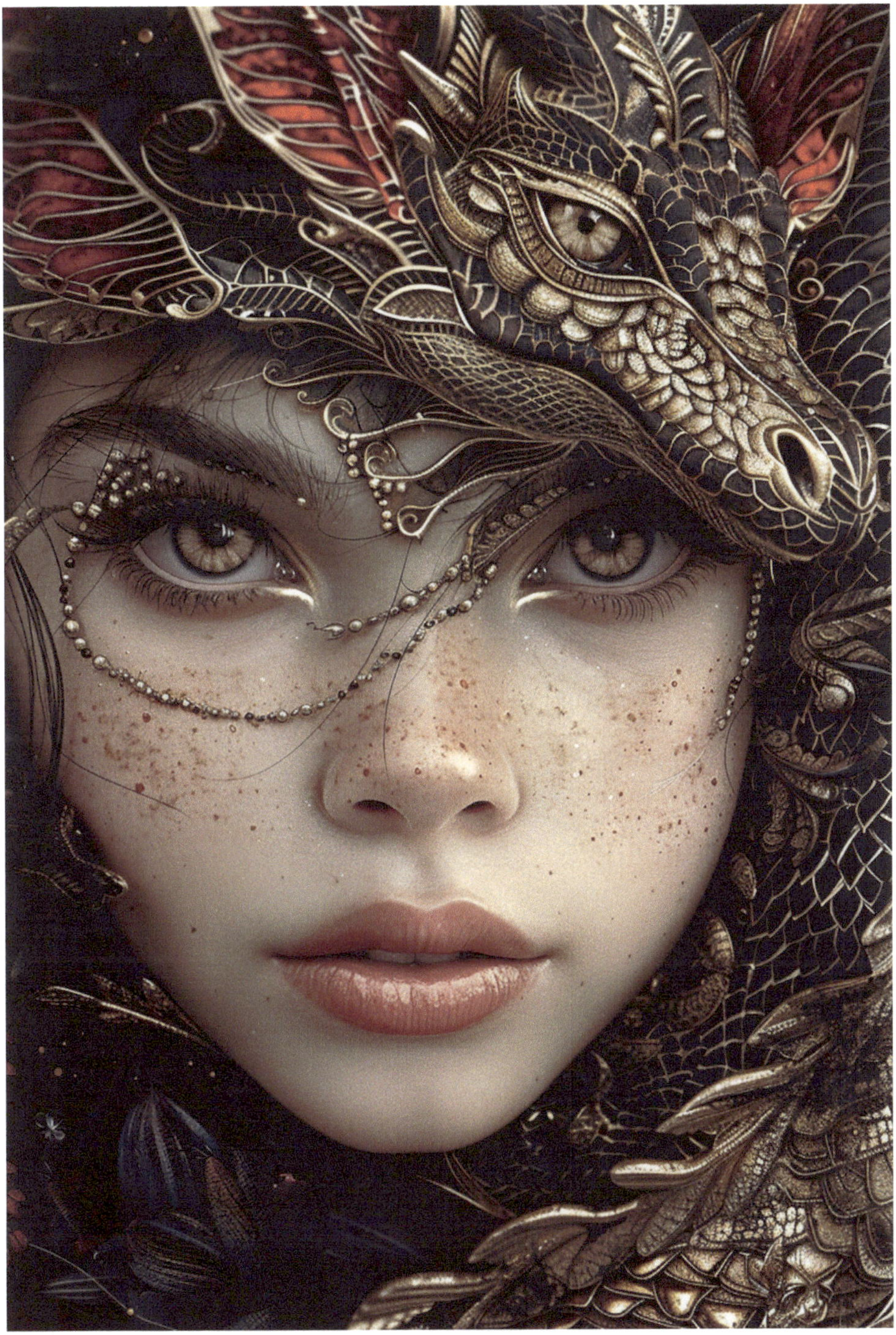

## Being

This image impacts me greatly. I sense inner sadness in this character, who I engage with as feminine. She is beautiful and intriguing. Her eyes are wide and crystalline yet not sparkling. I want to know her story — what is she? Where has she come from? What has happened to her? Of course, "she" may not be a being but a sculpture that is inviting engagement to explore the emotions and reactions the figure engenders.

The figure's mystical and otherworldly aura, the uniqueness and sense of isolation, and the artistic and imaginative details all contribute to a complex portrayal that resonates with a variety of archetypal themes.

- The exaggerated features, such as the oversized eyes and small nose, emphasize vulnerability and naivety. Her expression is neutral yet slightly melancholic, suggesting a sense of wonder or curiosity tinged with sadness. These align with **The Innocent/Child Archetype**.
- The earthy tones , intricate patterns and markings on her face and body, along with the ornate headpiece, suggest a connection to ancient or mystical traditions, invoking the **The Mystic/Seeker Archetype**.
- The figure's unconventional and exaggerated features and elaborate adornments could signify **The Outsider/Misfit Archetype**.
- The detailed craftsmanship of the sculpture, with its intricate patterns and textures, reflects a high level of artistic skill and creativity suggest the vivid imagination of **The Artist Archetype**.

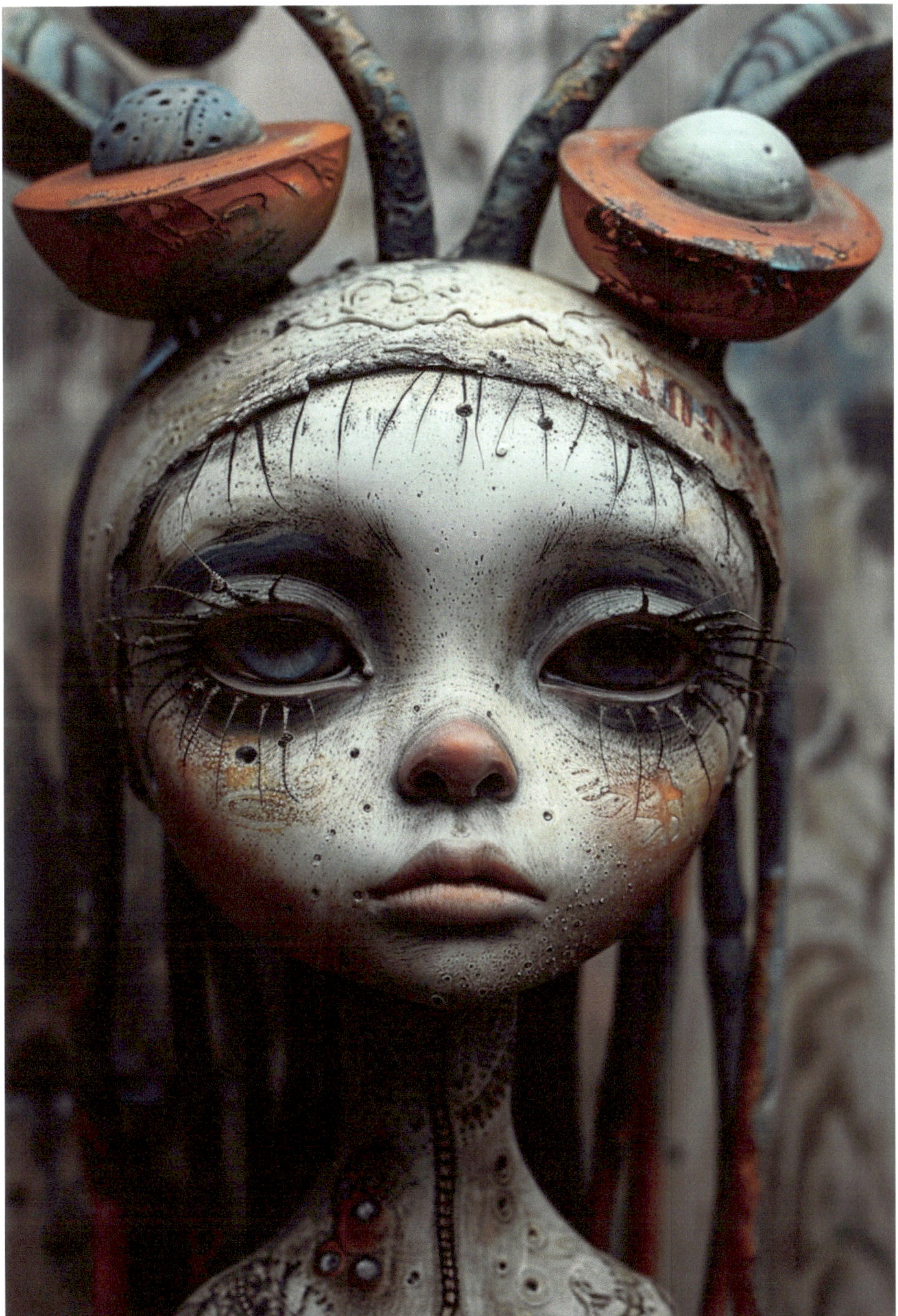

## Sad Robot

I sense a tragic sadness in this small, seemingly abandoned robot in a scene that evokes a sense of loneliness and decay. Is it lost? Has it given up hope? The image serves as a poignant reflection on the human condition, the struggle for connection, and the enduring spirit in the face of adversity and isolation.

- The robot's worn and weathered appearance, along with its solitary presence in an abandoned setting, suggests the feelings of neglect, vulnerability, and isolation of **The Abandoned Child Archetype**.
- The robot, marked with the number "123," may symbolize one of many similar entities, emphasizing its anonymity and lack of unique identity. **The Orphan Archetype** highlights the struggle to find a place in the world and the need to form meaningful connections.
- The robot's apparent lack of purpose or companionship underscores its status on fringes of society, either by choice or circumstance. **The Wanderer/Outcast Archetype** must navigate the world without the support and comfort of social structures.
- The robot's continued existence and presence amidst decay suggest the tenacious spirit and the potential for endurance and renewal of **The Dystopian Hero Archetype**.
- The robot's simple, almost childlike design, combined with its worn and abandoned state, can symbolize the **Lost Innocence Archetype**.
- The desolate urban environment, with its graffiti-covered walls and scattered debris, serves as a visual representation of **The Ruins as a Symbol of Decay**.

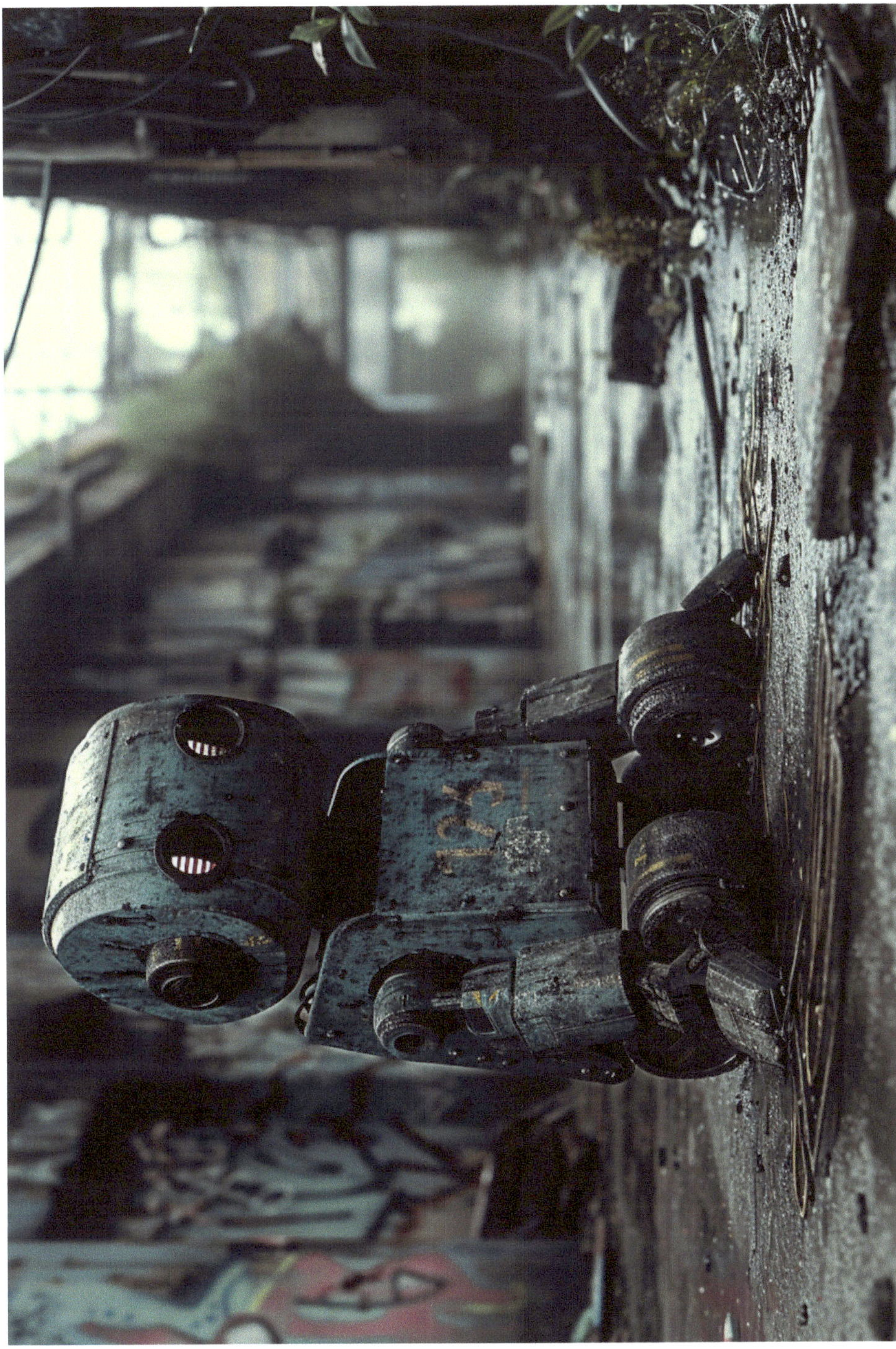

## Archetypographer

This image speaks to me of boldly, calmly, and carefully looking into the realm of thought. This person is mapping reality. This book concerns *archetypes* and woman depicted is practicing this as a way of mapping reality — an intellectual endeavor I have called ***Archetypography***.

- Her face overlaid with technological elements, such as circuitry and digital graphics, represents the fusion of organic and technological, and the intersection of humanity and technology, and so may involve a gentler aspect of **The Cyborg Archetype**.
- The glowing eye may represent a new form of vision or insight that enables advanced capabilities to perceive hidden truths or future possibilities as with **The Seer** and/or **Visionary Archetypes**.
- The **Creator** and **Inventor Archetypes** are suggested for providing the capacity for creativity and problem-solving, as well as the potential for technology to expand and augment these abilities.
- **The Trickster Archetype** embodies both positive and negative aspects, such as the potential for technology to disrupt and transform society, as well as the risks and unintended consequences of technological advancement.
- The image's blend of human and technological elements symbolizes the evolution of the **Collective Unconscious in the Digital Age**.

The artwork serves as a visual representation of the complex and evolving relationship between humanity and technology, and the potential for technological advancement to reshape our identities, experiences, and cultural narratives.

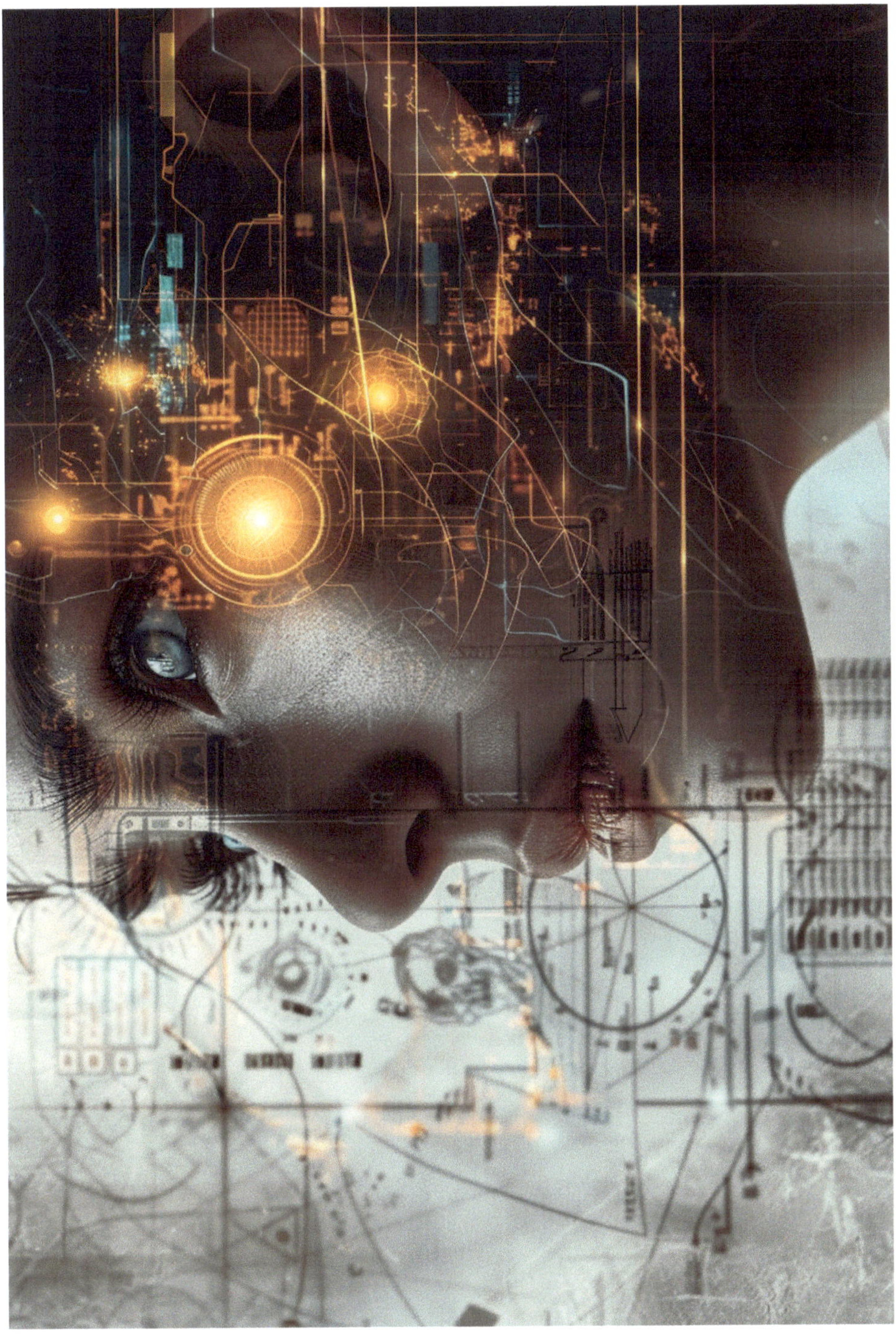

## Dejected Girlbot

I fill with empathy, an abject sadness, am terribly troubled by this beautiful Girlbot's situation. Her dejection is evident and this suggests that the loss of her left food is, as with my sadness, unaddressable. Has she been left behind? Where are her people. She seems without hope.

This child-like figure with robotic or cyborg enhancements, sits on a wooden structure in a natural setting. The juxtaposition of youth, technology, and nature invites an exploration of several human archetypes and themes:

- **The Divine Child Archetype** represents innocence, potential, and the promise of growth and renewal.
- **The Cyborg Archetype** explores the integration of technology with the human body and the blurring of lines between the organic and the artificial.
- **The Orphan/Wanderer Archetype** represents independence, self-discovery, and the search for belonging.
- **The Nature vs. Technology Duality** raises questions about the relationship between humanity and the environment, the role of technology in society, and the potential for harmony or conflict between the natural and the artificial.
- The image's blend of human and robotic elements evokes the concept of **The Uncanny Valley** which refers to the unsettling feeling that arises when humanoid objects or characters appear almost, but not quite, human.

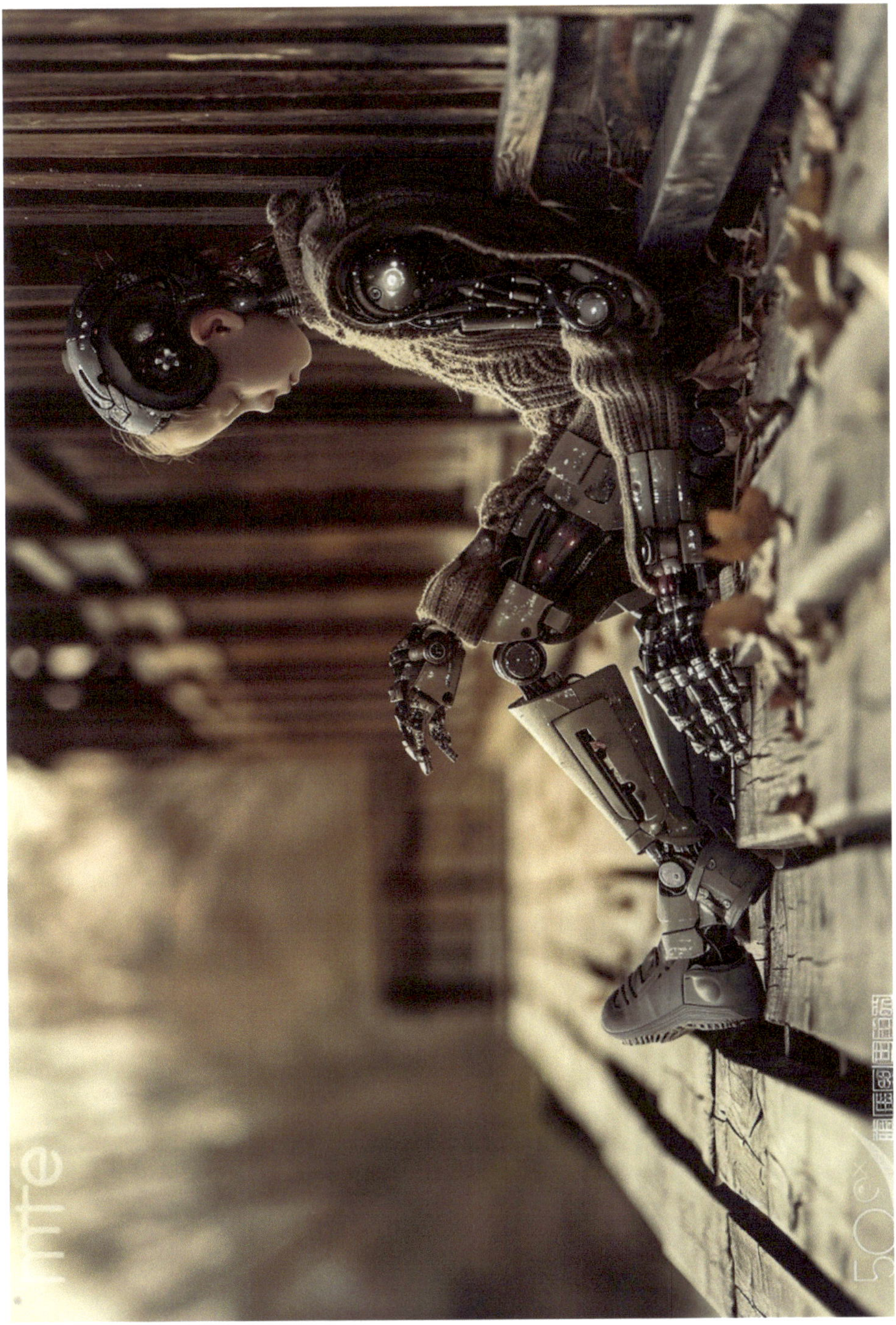

## Peaceful Contemplation

I am immersed in peacefulness and stillness, yet remain actively aware. Draped by the beauty of existence, I am one with nature. The luscious mysteries of moon and water connect the indomitable Self, the patient but ever inquisitive Spirit, and strength from the depths of the Psyche.

This image depicts a serene, mystical scene with a figure sitting by a body of water under a glowing moon. The figure appears contemplative and is surrounded by nature, evoking several human archetypes and themes:

- The figure in the image embodies **The Mystic/Seeker Archetype**, which represents a quest for spiritual knowledge, enlightenment, and a deeper understanding of the self and the universe.
- The prominent, glowing moon in the image can symbolize **The Moon as a Feminine Archetype**, often associated with intuition, emotion, and the cycles of life.
- The solitary nature of the figure and the tranquil, isolated setting can evoke **The Hermit Archetype**, which represents withdrawal from the world, introspection, and the search for inner wisdom.
- The lush, vibrant natural setting, with its flowers and serene body of water, connects the image to **The Nature Archetype**.
- The body of water in the image can represent the unconscious mind, a vast and often mysterious realm of emotions, memories, and hidden thoughts. **The Water as the Unconscious Mind** symbolizes a willingness to explore and engage with the depths of the psyche.

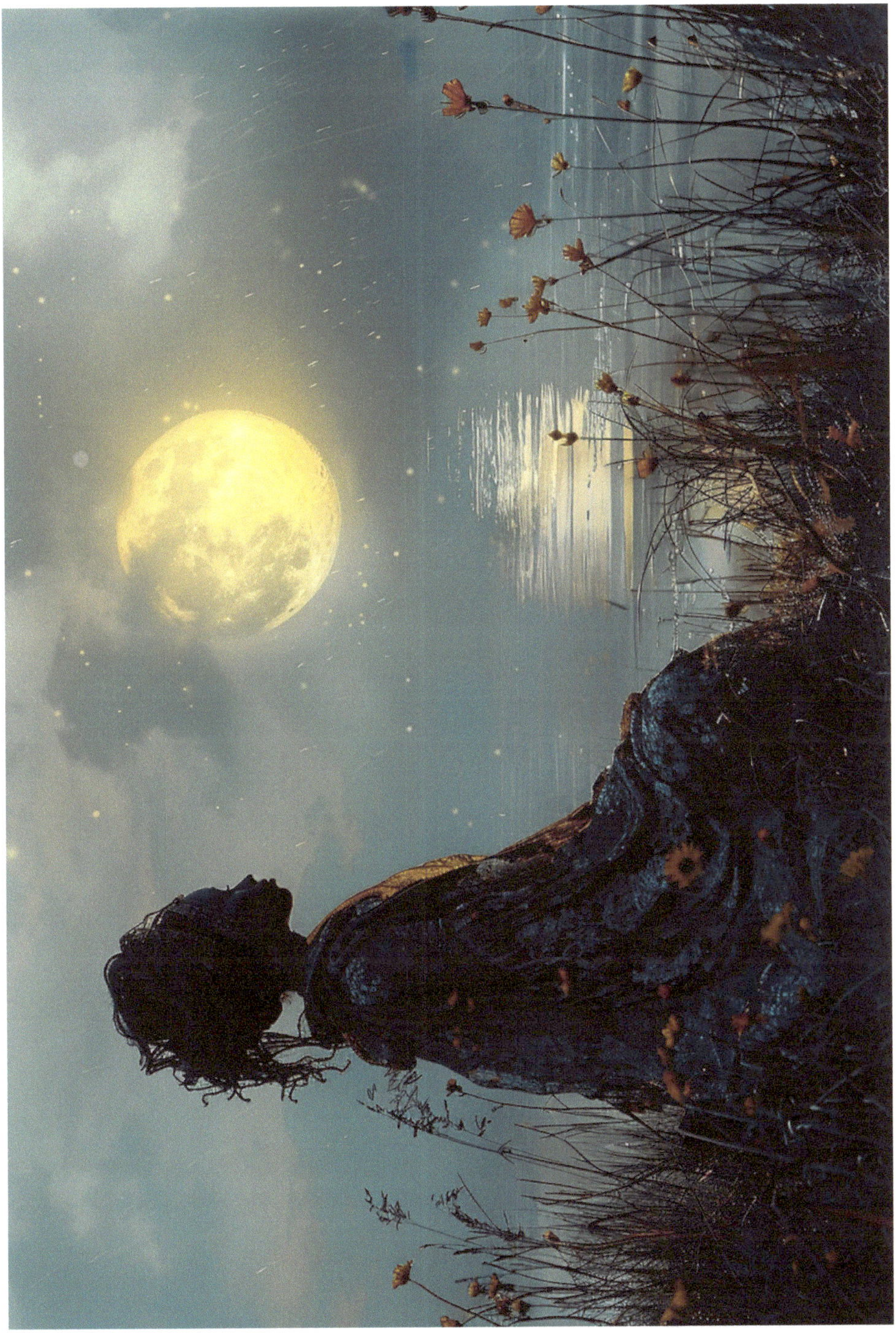

## Empire of Tomorrow

I am captivated by the powerful scene which evokes bravery, prowess, and gumption in the face of minuteness, staggering odds, and sheer isolation in a barren and hostile foreign environment. The spaceship is a testament to human ingenuity and the willingness of society to explore.

This image depicts a solitary figure in a red spacesuit standing in a desolate, otherworldly landscape, facing a towering structure that resembles a lighthouse or beacon. The scene evokes a sense of isolation, mystery, and exploration, inviting analysis through the lens of several human archetypes:

The central figure in the spacesuit embodies the **The Explorer/Seeker Archetype**, which represents the desire to discover, learn, and understand the unknown.

The image can be seen as a visual representation of **The Hero's Journey** in which the protagonist ventures into the unknown, faces challenges, and ultimately returns transformed.

The towering structure in the image resembles a lighthouse or beacon, which can symbolize the guidance, hope, and illumination of **The Beacon/Lighthouse as a Symbol of Guidance**.

The artwork serves as a visual representation of the human desire for exploration, discovery, and personal growth, as well as the challenges and mysteries that arise from venturing into the unknown.

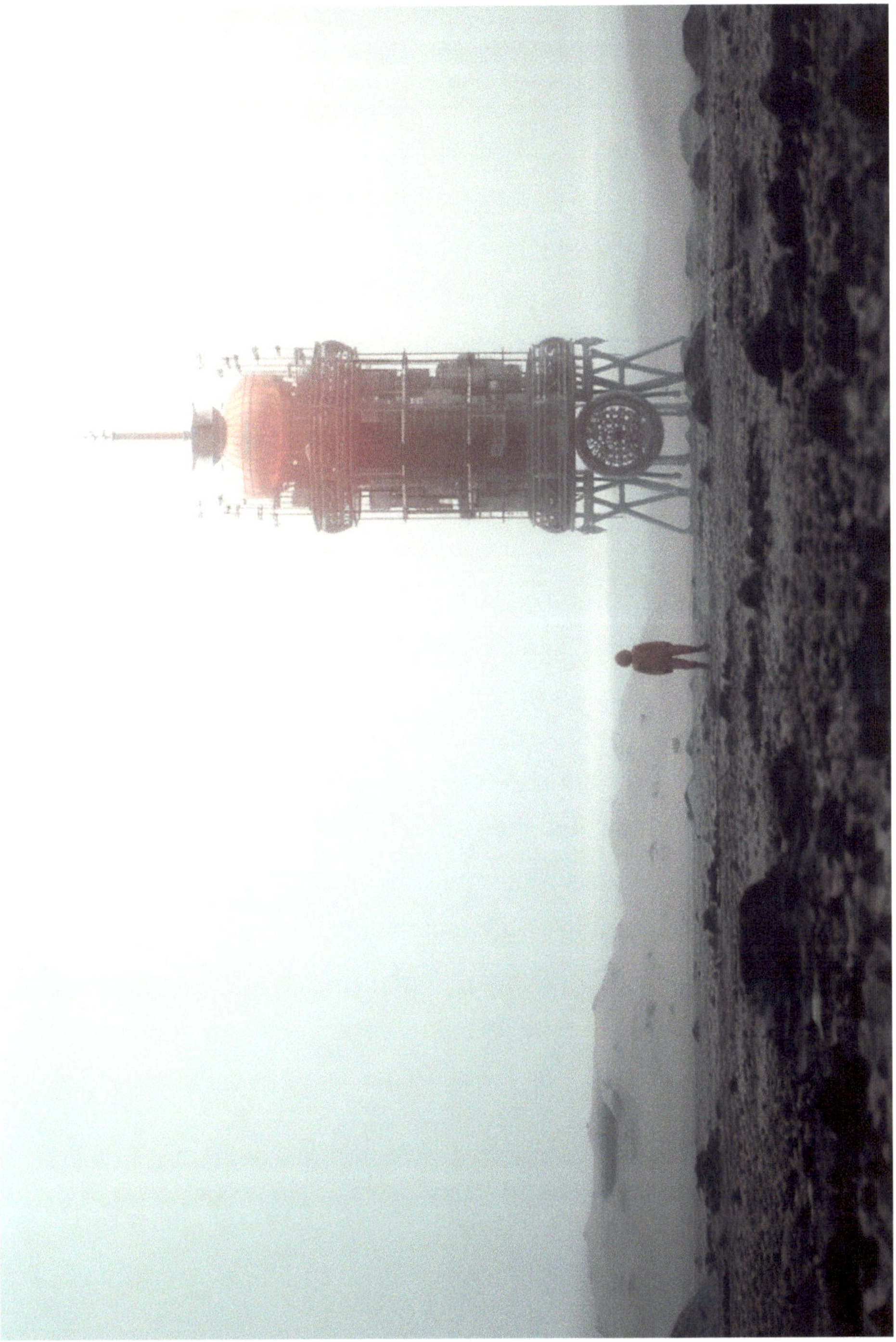

## The Fabric of Reality

Have I been here? This place feels so familiar. Has it been in a dream? Perhaps this place is an archetype common to human beings from before the paving-over of the world. Is it cold there? Or warm? Do I have sensation? I have been traveling and have come upon this place. It appears to be abandoned and dilapidated. Why is that? The world has moved on. These remote places that people once inhabited are forgotten. We stumble upon them and are reminded of our roots. Maybe.

NOTE — At first, I thought that Mistral.ai had "hallucinated" on the interpretation of this image. It said, "This image depicts a person standing in a contemplative pose amidst a vast, open landscape with a prominent, ancient-looking tree in the background." Wow, yes, such a person is here - it is me, the Viewer!

- The **The Seeker/Wanderer Archetype** represents a journey of self-discovery, exploration, and the search for meaning or truth.
- **The Sage/Philosopher Archetype** symbolizes wisdom, introspection, and a deep understanding of life and the universe.
- The prominent, ancient-looking tree in the background can symbolize **The Tree of Life Archetype** which represents growth, renewal, and the interconnectedness of all living things.

The image may represent a moment of profound insight or contemplation, as the individual seeks to grasp the essence of existence and their place within it.

NOTE — Indeed, the image did represent such a moment!

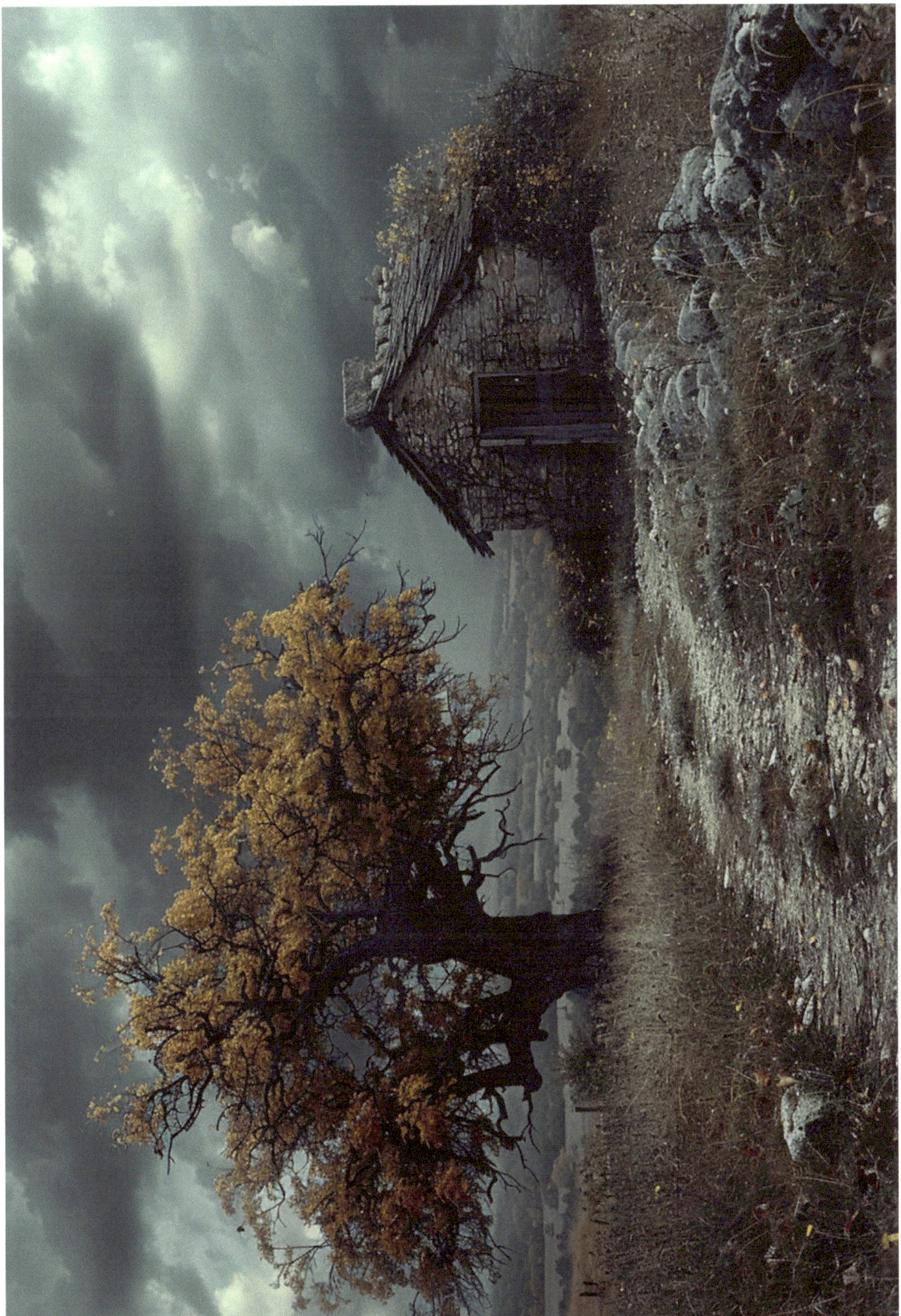

## Tycoon

I do not want to need to stare down this man. I don't know if I could do so. His self-confidence is complete. I aspire to such. His hair and beard are manicured to perfection, and yet it looks an effortless attainment. His deeply focused eyes are piercing through the tinted eyeglasses that perfectly frame his face and we know his attention on us is undivided.

This image portrays a man with distinct features that can be analyzed through the lens of human archetypes, which are universal, symbolic patterns of behavior or roles that recur in stories and cultures across the world.

The man's mature appearance, intellectual demeanor, mysterious aura, non-conformist style, and artistic aesthetic all contribute to a complex and intriguing portrayal that resonates with multiple archetypal themes.

- His serious expression and the slight furrowing of his brow indicate deep thought or concern, which are traits often associated with **The Sage/Mentor Archetype**.
- The moody, dim lighting and the cool color tones create a sense of mystery and introspection. His gaze is direct and intense, which might imply a search for deeper meaning or understanding. These reflect **The Mystic/Seeker Archetype**.
- The overall aesthetic of the image, with its dramatic lighting and composition, and the attention to detail in his appearance suggest **The Creator/Artist Archetype**.

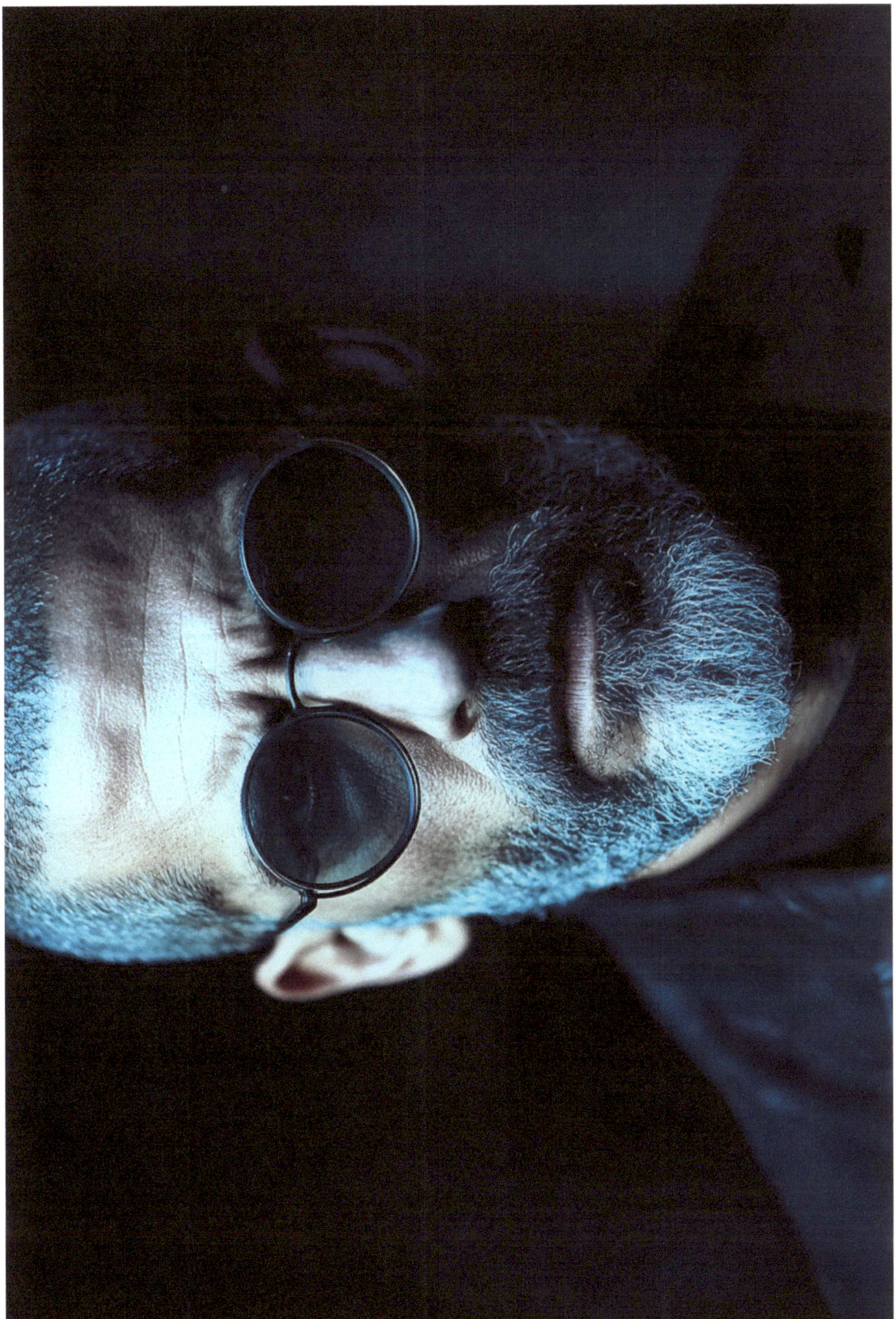

# GALLERY 2

## About this Gallery

This set of images is presented each as a four page spread. As with first set, the lead spread has text on the left and the image on the right.

The first paragraph of text is written by me and conveys my reaction to the image. The title for each image is provided by me (and was chosen for simplicity and essence).

The subsequent paragraph is the summary provided by **Mistral.ai**.

A second two page spread is an assessment of the image by the **Mistral.ai** system.

The prompt I used with **Mistral.ai** to assess the images was:

> *Please provide an analysis of the following image exploring and assessing these attributes:*
>
> - *Color Palate*
> - *Concept*
> - *Composition*
> - *Creativity*
> - *Emotional Impact*
> - *Intention*
> - *Style*

## Arjuna

### Stuart's Reaction

My sense is that this is an image of Arjuna. According to Google's Gemini AI:

> *Arjuna is a character in the Hindu epic Mahābhārata and the philosophical dialog Bhaghdad Gita. He is a warrior and archer who is a central figure in the Kurukshetra War.*

But… whether this is Arjuna or an archetypal similarity, the image captivates my mind and eye. I am beholden to the lyrical qualities inherent in the poise, suspended in air and also perhaps in time, assessing for action, or… for non-action. The suffuse color palate is sweeping, drawing the eye's attention to the face, which is subtly yet clearly illuminated from a light source that casts shadows on the scene.

### Mistral.ai's Summary

This image is a beautifully executed blend of traditional and modern styles, depicting a divine figure with rich symbolism and emotional impact. The use of vibrant colors, dynamic composition, and intricate details showcases the artist's creativity and skill, resulting in a visually and emotionally engaging piece.

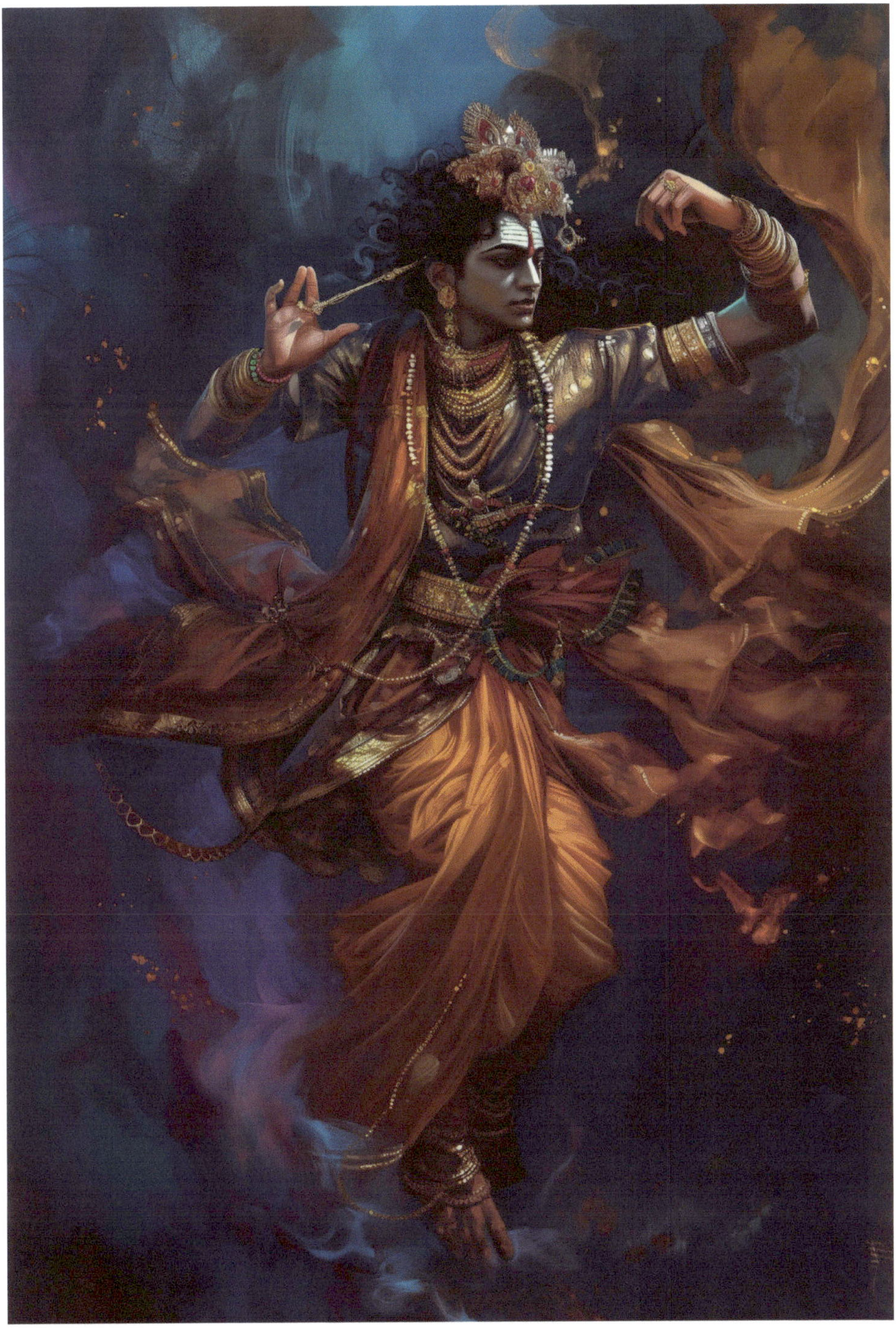

## Mistral.ai's Assessment

- **Color Palate:** The color palette is rich and vibrant, dominated by warm hues of orange, gold, and red, contrasted with cooler tones of blue and purple. The warm colors are used for the figure's attire and accessories, creating a sense of energy, vitality, and spiritual significance. The cooler tones in the background provide a dramatic contrast, enhancing the figure's prominence and adding depth to the scene. The use of gold accents highlights the divine and regal nature of the figure.
- **Concept:** The concept appears to be a depiction of a divine or mythological figure, likely inspired by Hindu deities such as Krishna or Vishnu. The figure is adorned with traditional ornaments, including a crown, jewelry, and a flute, which are symbolic of divine attributes and spiritual power. The dynamic pose and flowing garments suggest movement and energy, possibly representing a dance or a divine performance.
- **Composition:** The composition is dynamic and balanced, with the figure centrally placed and occupying most of the frame. The use of diagonal lines and flowing fabrics creates a sense of movement and energy. The background is abstract and swirling, which draws the viewer's attention to the figure and enhances the overall dramatic effect. The placement of the figure's arms and the direction of the gaze add to the sense of motion and engagement.
- **Creativity:** The creativity in this image is evident in the unique interpretation of a traditional mythological figure. The artist has blended traditional elements with a modern, dynamic style, using

vibrant colors and abstract backgrounds to create a visually striking and emotionally engaging piece. The use of light and shadow, along with the intricate details of the figure's attire and accessories, showcases the artist's skill and imagination.

- **Emotional Impact:** The emotional impact of the image is powerful and evocative. The figure's pose and expression convey a sense of divine grace, power, and serenity. The vibrant colors and dynamic composition create a feeling of energy and movement, while the abstract background adds a sense of mystery and otherworldliness. The overall effect is one of awe and reverence, inviting the viewer to contemplate the divine and spiritual themes represented.
- **Intention:** The intention behind this image seems to be a celebration of divine and spiritual themes, possibly inspired by Hindu mythology. The artist aims to capture the essence of a divine figure, highlighting attributes such as grace, power, and spiritual significance. The use of vibrant colors, dynamic composition, and intricate details serves to emphasize the figure's divine nature and create a visually and emotionally engaging piece.
- **Style:** The style of the image is a blend of traditional and modern elements. The figure's attire and accessories are inspired by traditional Hindu iconography, while the use of vibrant colors, abstract backgrounds, and dynamic composition gives the image a contemporary and stylized look. The artist has skillfully combined these elements to create a unique and visually striking interpretation of a mythological figure.

## Classic Seen

### Stuart's Reaction

Have I been here? No. Would I like to be there? Of Course! Can I be? Not a chance. This place has never yet been. This, to me, is a street that Havana should have had. Oh! could that be a street in a European town — one that exists only at the intersection of dream and imagination. This image is presents a photorealistic view of something that has not been seen. A classic scene with a classic car that has not been made. The scene is a fabrication that feels all so real. The scene is a luscious, the street has been constructed by the most meritorious of craftsman, and the features are an imprecise amalgam of elements dear to human beings that travel to foreign places for a touching experience that is both novel and familiar. Welcome to the hitherto fore unseen.

### Mistral.ai's Summary

The image is a beautifully composed and emotionally resonant piece that effectively uses color, composition, and creativity to convey a sense of nostalgia, warmth, and tranquility. It invites the viewer to appreciate the beauty in everyday moments and evokes a sense of longing for simpler times.

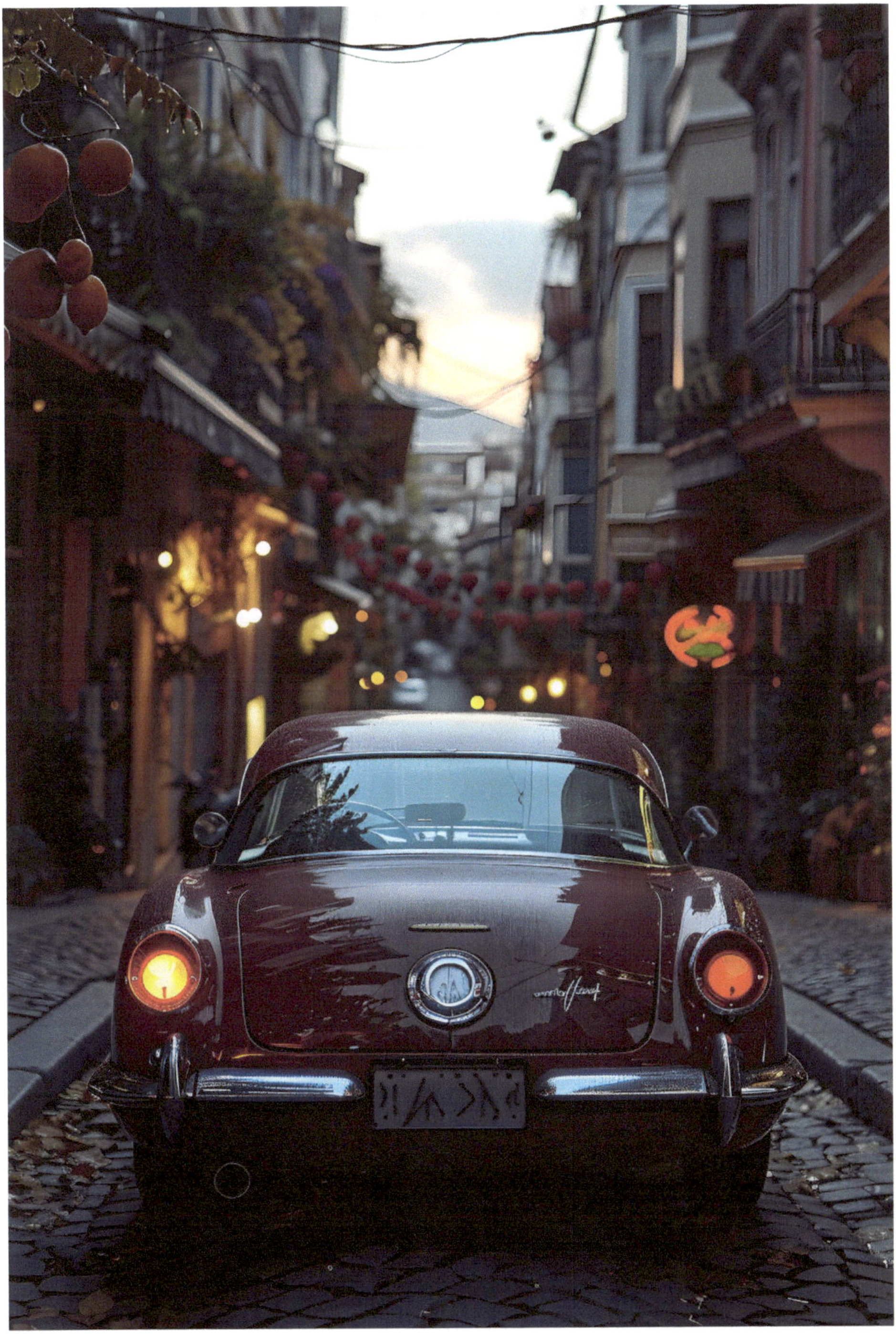

## Mistral.ai's Assessment

- **Color Palate:** The image features a warm and inviting color palette. The red car in the foreground is the most vibrant element, drawing the viewer's attention. The surrounding environment is adorned with warm hues of orange, yellow, and brown, creating a cozy and nostalgic atmosphere. The soft lighting from the lanterns and the natural light in the background add depth and warmth to the scene.
- **Concept:** The concept of the image seems to revolve around nostalgia, charm, and the beauty of everyday life. The vintage car, the cobblestone street, and the quaint buildings with their decorative elements evoke a sense of a bygone era. The image captures a moment of tranquility and simplicity, inviting the viewer to appreciate the beauty in ordinary moments.
- **Composition:** The composition is well-balanced, with the red car serving as the focal point. The car is centrally placed, drawing the viewer's eye immediately. The leading lines of the cobblestone street and the arrangement of the buildings guide the viewer's gaze deeper into the scene. The decorative elements, such as the lanterns and plants, add layers of interest without overwhelming the main subject.
- **Creativity:** The image is creative in its use of lighting, color, and composition to create a sense of warmth and nostalgia. The vintage car and the charming street scene are not just visually appealing but also tell a story of a place and time that feels both familiar and magical. The attention to detail in the decorations and the overall atmosphere adds to the creative appeal.

- **Emotional Impact:** The image evokes feelings of warmth, nostalgia, and tranquility. The vintage car and the quaint street scene create a sense of longing for simpler times. The soft lighting and warm colors contribute to a cozy and inviting atmosphere, making the viewer feel a sense of comfort and peace. The overall mood is one of serenity and appreciation for the beauty in everyday life.
- **Intention:** he intention behind the image appears to be to capture a moment of beauty and tranquility in an everyday setting. It aims to evoke feelings of nostalgia and appreciation for the simple pleasures in life. The careful composition and attention to detail suggest an intention to create a visually appealing and emotionally resonant scene.
- **Style:** The style of the image is nostalgic and romantic, with a touch of vintage charm. The use of warm colors, soft lighting, and vintage elements contributes to a timeless and classic aesthetic. The attention to detail in the decorations and the overall atmosphere adds to the stylistic appeal, making the image both visually striking and emotionally engaging.

# Ice Stare

Stuart's Reaction

Oh my, this image bites at the ice and frost, her intensity raises the temperature of her skin and my blood. There is raw power here that is held calmly at rest. The judicious use of red on lips and below the eye have these stand in stark juxtaposition to the chilling and muted frigidity of the environment. Her eye is clear, and in the iris we see the sparkle of the universe around the black hole of the pupil, within which lies a woman of confident capability, of refined femininity and indescribable intelligence. Her lips are parted — does she speak? Is she waiting to speak? Do I dare speak first?

**Mistral.ai's Summary**

The image is a powerful and emotionally resonant piece that effectively uses color, composition, and creativity to convey a sense of mystery, intensity, and emotional depth. It invites the viewer to explore the hidden layers of the subject's emotions and appreciate the beauty in its complexity.

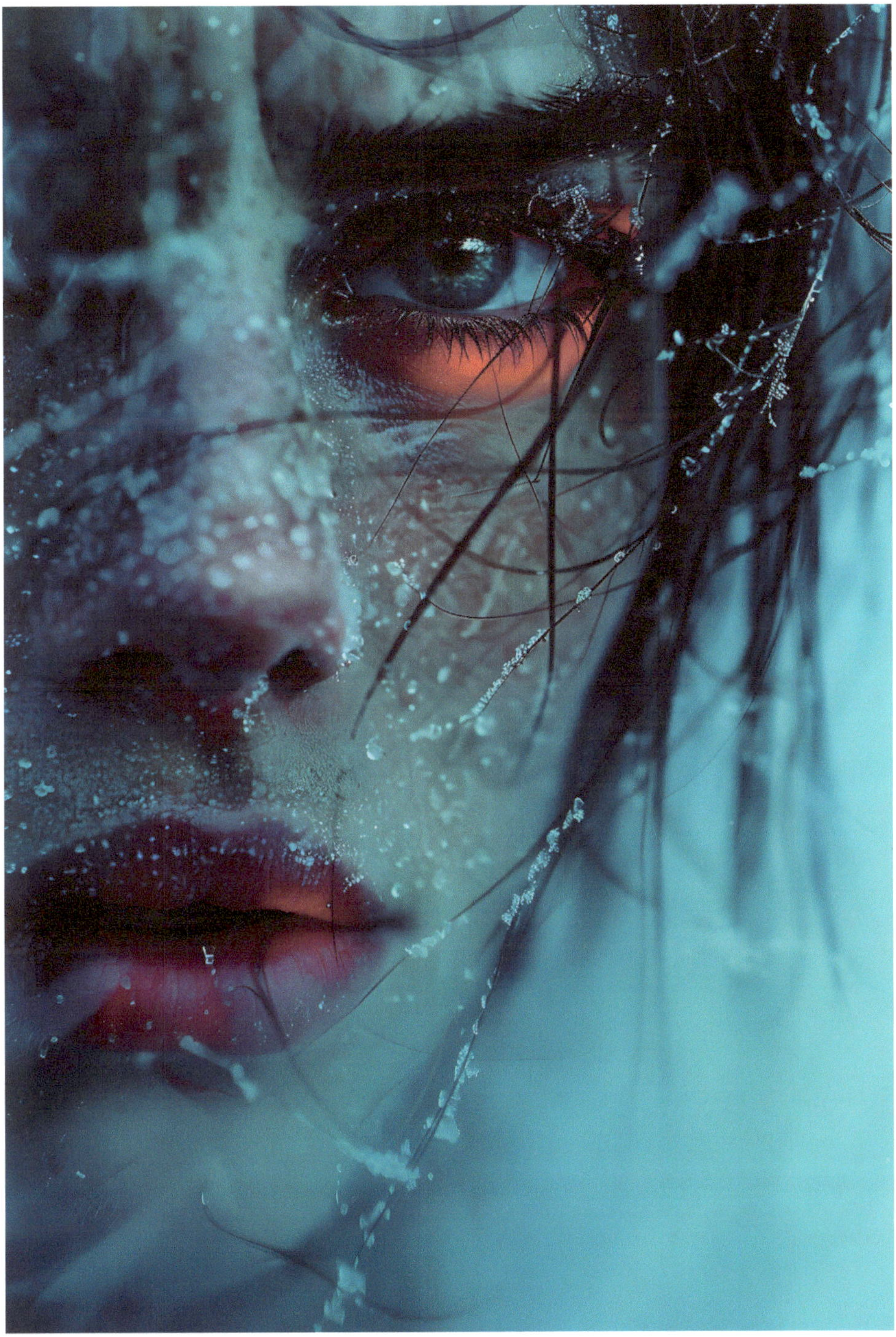

## Mistral.ai's Assessment

- **Color Palate:** The image features a striking contrast between cool and warm tones. The predominant cool blue and green hues create a sense of mystery and intrigue, while the warm red accents on the lips and around the eye add a dramatic and intense focal point. The interplay between these colors enhances the emotional depth and visual impact of the image.
- **Concept:** The concept of the image appears to be centered around themes of mystery, emotion, and introspection. The close-up of the face, partially obscured by a wet, textured surface, suggests a sense of vulnerability and hidden emotions. The vivid red accents on the lips and eye could symbolize passion, intensity, or inner turmoil.
- **Composition:** The composition is intimate and focused, with the face taking up most of the frame. The close-up perspective draws the viewer's attention to the details of the face, particularly the eye and lips. The wet, textured surface adds a layer of complexity and depth, creating a sense of separation between the viewer and the subject. The diagonal lines formed by the strands of hair add dynamic movement to the composition.
- **Creativity:** The image is highly creative in its use of color, texture, and composition to convey a sense of mystery and emotional depth. The wet, textured surface is an innovative element that adds a unique visual interest and symbolic meaning. The bold use of red accents against the cool background creates a striking contrast that enhances the overall impact of the image.

- **Emotional Impact:** The image evokes a range of emotions, including intrigue, curiosity, and a sense of hidden intensity. The cool tones and the wet, textured surface create a sense of distance and mystery, while the warm red accents suggest underlying passion or turmoil. The overall mood is one of introspection and emotional complexity.
- **Intention:** The intention behind the image seems to be to explore themes of emotion, mystery, and introspection. It aims to provoke the viewer to look beyond the surface and consider the hidden depths of the subject's emotions. The use of color, texture, and composition is deliberate and intended to create a visually striking and emotionally resonant piece.
- **Style:** The style of the image is contemporary and artistic, with a strong emphasis on color and texture. The use of bold contrasts and innovative elements, such as the wet, textured surface, contributes to a unique and captivating visual language. The overall aesthetic is dramatic and emotionally charged, making the image both visually striking and thought-provoking.

## Sculpture

### Stuart's Reaction

I could circle this sculpture entranced. I could sit on the floor beside it and meditate. I could stand by the window looking out yet be unable to unknown what is behind me. I want to rove my hand across the smoothly finished but unpolished surface. I want to explore the lighting. I want to know how was this fabricated? By who? And for what reason? Am I to ponder on the infinite or the number 8? Am I to admire the precision of the shape as well as the imprecision in the grain and the imperfections in the wood itself. All of this — perfection, subjectivity, and abstraction — come together in the focal point of a room that is the perfect container for the sculpture, itself set on a platform of ideal stone atop flooring masterfully lain. And so around and around we go.

### Mistral.ai's Summary

The image presents a beautifully crafted sculpture that combines a rich color palette, a thoughtful concept, and a well-balanced composition to create a piece that is both visually striking and emotionally resonant. The creativity and intention behind the sculpture are evident in its design and execution, making it a compelling work of art.

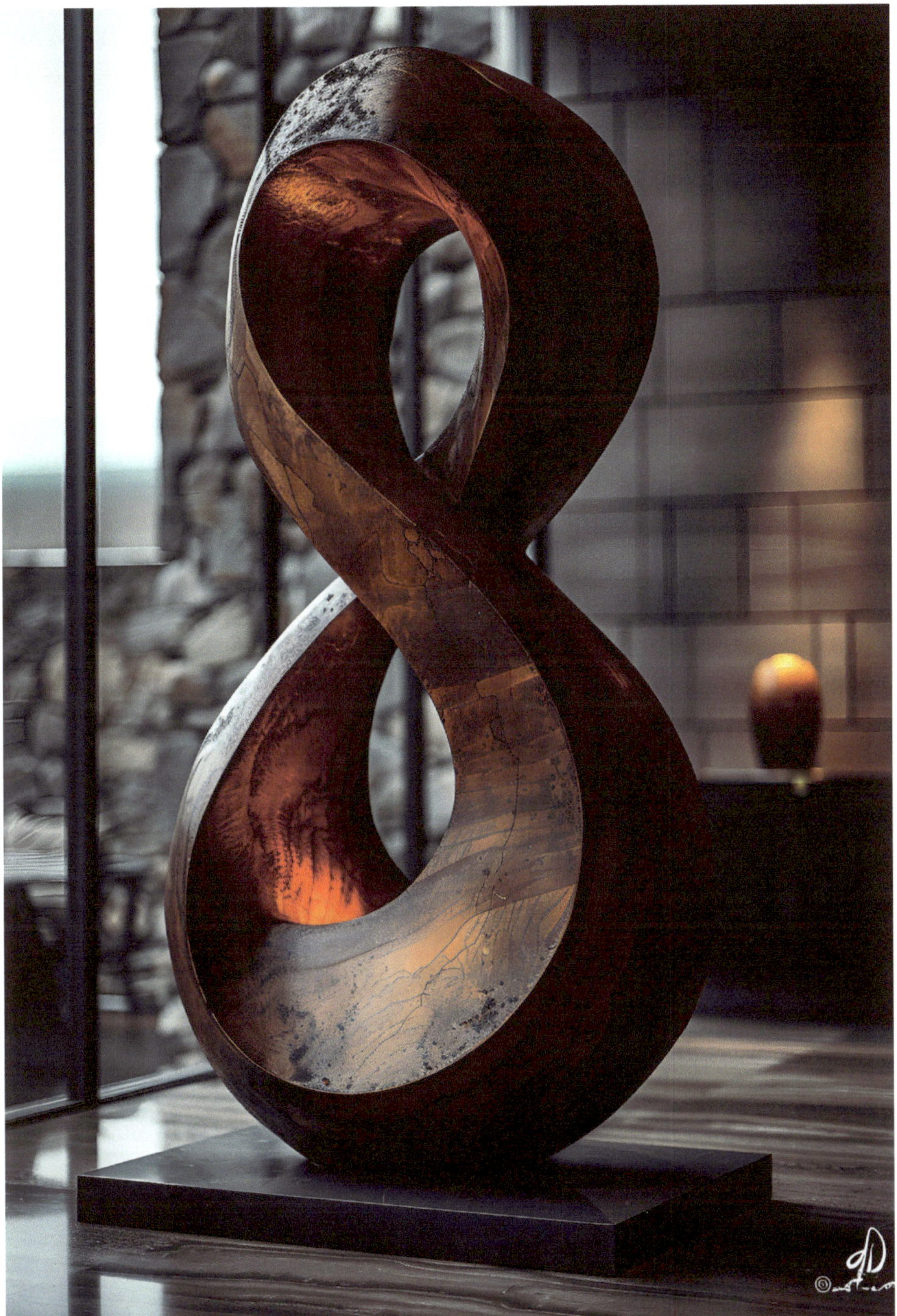

## Mistral.ai's Assessment

- **Color Palate:** The image features a rich, warm color palette dominated by earthy tones. The sculpture itself is a blend of deep browns and rusty oranges, which give it a natural, almost organic feel. The warm lighting from within the sculpture enhances these tones, creating a sense of depth and warmth. The background is more muted, with cooler grays and blacks, which contrasts nicely with the sculpture and makes it stand out.
- **Concept:** The concept of the sculpture appears to be centered around the idea of infinity or continuity. The shape of the sculpture, resembling an infinity symbol or a Möbius strip, suggests themes of endlessness, unity, and the cyclical nature of life. The interplay of light and shadow within the sculpture further emphasizes these themes, creating a sense of movement and flow.
- **Composition:** The composition is well-balanced, with the sculpture placed centrally in the frame. The use of lighting is particularly effective, as the warm glow from within the sculpture draws the viewer's eye inward, creating a focal point. The background elements, such as the stone wall and the blurred lights, provide context without distracting from the main subject. The reflection on the floor adds depth and symmetry to the composition.
- **Creativity:** The sculpture itself is a testament to creativity, with its unique shape and the innovative use of light. The artist has taken a simple concept—the infinity symbol—and transformed it into a visually striking piece. The combination of materials and the play of light and shadow add layers of complexity and intrigue.

- **Emotional Impact:** The sculpture evokes a sense of calm and contemplation. The warm, inviting light and the smooth, flowing lines of the sculpture create a soothing effect. The themes of infinity and continuity can inspire feelings of hope, resilience, and the interconnectedness of all things. The overall atmosphere is one of serenity and introspection.
- **Intention:** The intention behind the sculpture seems to be to explore the concept of infinity and the interconnectedness of life. The use of light and the choice of materials suggest a desire to create a piece that is both visually appealing and thought-provoking. The sculpture invites viewers to reflect on the cyclical nature of existence and the beauty of continuity.
- **Style:** The style of the sculpture is modern and abstract, with a focus on form and material. The use of light as an integral part of the sculpture is a contemporary touch that adds a dynamic element to the piece. The smooth, polished surface of the sculpture contrasts with the rough texture of the stone wall in the background, creating a visually interesting juxtaposition.

## Uncertain Doorway

### Stuart's Reaction

I shiver with fear. Cold dread fills my veins. I have been here in dreams I don't recollect. I have stood before this doorway in the anteroom of anticipation. Dare I discover what is on the on the other side. My soul shivers at the thought. Is the door locked? Will the knob turn? But wait — am I on the inside? A prisoner? Do I have the gumption to escape? The red lanterns call, the split red paint suggest blood, and the red ideograms command notice but I am unable to infer any meaning. Death has preceded me here. Or has it invited me here. I must get away, and yet know that to live fully and completely, I must have the courage to face the uncertainty that is presented to me.

### Mistral.ai's Summary

The image is a beautifully crafted piece that combines a muted color palette, a thoughtful concept, and a well-balanced composition to create a scene that is both visually striking and emotionally resonant. The creativity and intention behind the image are evident in its design and execution, making it a compelling work of art.

Uncertain Doorway

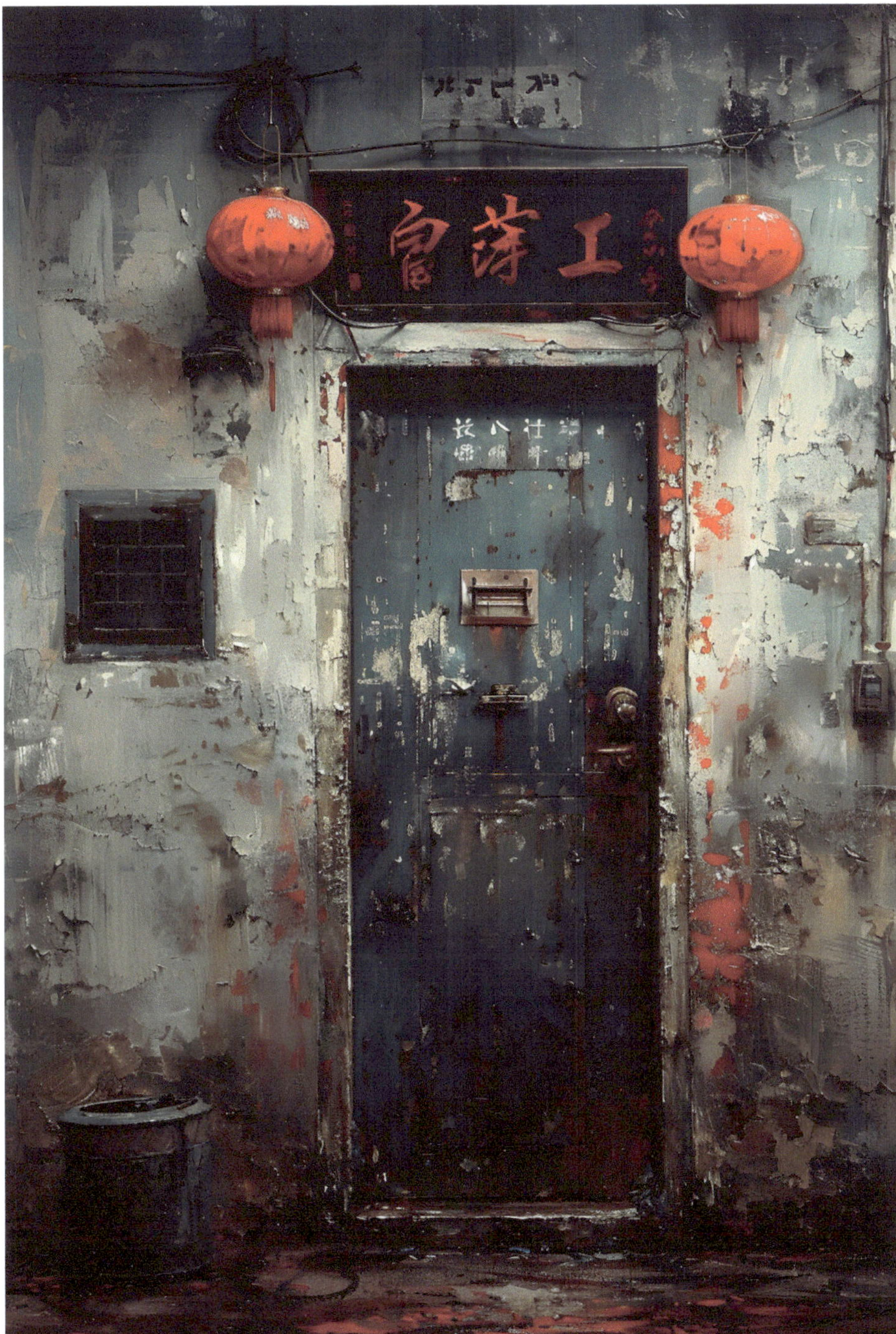

## Mistral.ai's Assessment

- **Color Palate:** The image features a muted and somewhat desaturated color palette, dominated by shades of blue, gray, and brown. The worn-out appearance of the walls and door, with peeling paint and rust, adds to the overall sense of age and decay. The bright red lanterns and Chinese characters provide a stark contrast, drawing the viewer's attention and adding a touch of vibrancy to the otherwise subdued scene.
- **Concept:** The concept of the image appears to revolve around themes of decay, nostalgia, and cultural heritage. The dilapidated state of the building suggests a sense of abandonment or neglect, while the traditional elements like the red lanterns and Chinese characters hint at a rich cultural past. The juxtaposition of these elements creates a narrative of a place that was once vibrant but has since fallen into disrepair.
- **Composition:** The composition is balanced, with the door serving as the central focal point. The red lanterns and Chinese characters above the door draw the eye upward, creating a sense of vertical movement. The worn-out walls and the small window to the left of the door add depth and texture to the scene. The trash can and the electrical box on the right side provide additional context and balance the composition.
- **Creativity:** The image showcases creativity in its use of contrast and texture. The juxtaposition of the bright red lanterns against the muted, decaying background is visually striking. The detailed textures of the peeling paint and rust add a tactile quality to the image, making it feel

almost palpable. The inclusion of cultural elements like the lanterns and Chinese characters adds a layer of depth and meaning.

- **Emotional Impact:** The image evokes a sense of melancholy and nostalgia. The decaying state of the building suggests a loss of vibrancy and life, while the cultural elements hint at a rich past. The overall atmosphere is one of quiet reflection, inviting the viewer to ponder the passage of time and the transience of all things. The red lanterns, traditionally symbols of good fortune and joy, add a touch of hope and warmth to the scene.
- **Intention:** The intention behind the image seems to be to explore themes of decay, nostalgia, and cultural heritage. The artist likely aims to evoke a sense of the passage of time and the transience of life, using the contrast between the vibrant cultural elements and the decaying background to create a poignant narrative. The image invites viewers to reflect on the past and the inevitable changes that come with time.
- **Style:** The style of the image is realistic with a touch of impressionism. The detailed textures and the play of light and shadow create a sense of depth and realism, while the muted color palette and the emphasis on texture give the image a somewhat painterly quality. The inclusion of cultural elements adds a layer of symbolism and meaning, making the image both visually appealing and thought-provoking.

## Stairway

### Stuart's Reaction

Again, the intense red. Just often enough images include this powerful element. Google informs me that:

> *Dark red symbolizes passion and desire. Its deep hue also suggests mystery and suspense. Dark red carries connotations of luxury, wealth, and strength.*

Indeed, these attributes seem to swim in this scene. Yes, these are homes of wealthy dynasties. There is great mystery here as to what is within, the parlous especially so illuminated. The lanterns invite one to ascend. What is up there, hidden in the fog? Have we been invited to proceed? Or are we trespassing if we dare to proceed? I'd like to stand on one of the balconies and look out for I believe I'd feel like a Prince. And up on the right, written in white, simple letterforms that remind me that I am as yet unprepared to find a way in to explore libraries of the mind.

### Mistral.ai's Summary

The image is a beautifully crafted piece that combines a striking color palette, a thoughtful concept, and a dynamic composition to create a scene that is both visually striking and emotionally resonant. The creativity and intention behind the image are evident in its design and execution, making it a compelling work of art.

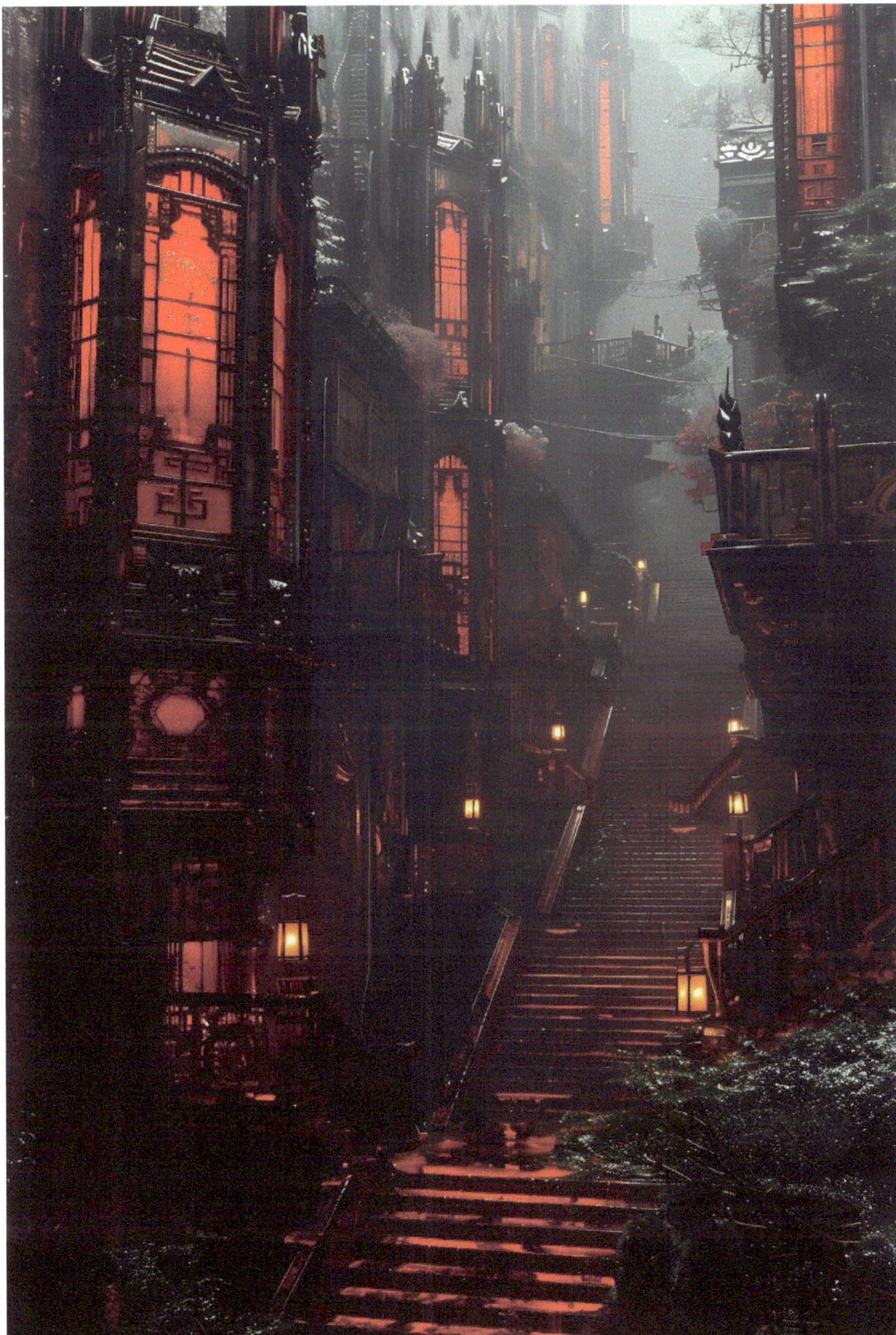

## Mistral.ai's Assessment

- **Color Palate:** The image features a striking contrast between warm and cool tones. The dominant colors are deep reds, oranges, and blacks, which create a sense of warmth and intensity. These warm tones are juxtaposed against cooler, muted grays and greens in the background, adding depth and balance to the scene. The use of red, in particular, draws the viewer's attention and evokes a sense of urgency and drama.
- **Concept:** The concept of the image appears to be a blend of traditional and futuristic elements, creating a sense of a dystopian or cyberpunk world. The architecture and design elements, such as the lanterns and staircases, suggest a traditional Asian influence, while the overall atmosphere and lighting hint at a futuristic, almost otherworldly setting. The image seems to explore themes of contrast, transformation, and the fusion of old and new.
- **Composition:** The composition is dynamic and complex, with multiple layers and focal points. The staircases and buildings create a sense of depth and perspective, drawing the viewer's eye into the scene. The use of light and shadow is particularly effective, with the bright red lanterns and illuminated windows serving as focal points that guide the viewer's gaze through the image. The misty, atmospheric background adds a sense of mystery and intrigue.
- **Creativity:** The image showcases a high level of creativity in its blend of traditional and futuristic elements. The intricate details of the architecture and the use of light and color create a visually striking and unique scene. The juxtaposition of traditional lanterns and

modern, almost alien-like structures adds a layer of complexity and intrigue. The overall atmosphere is both familiar and otherworldly, inviting the viewer to explore and interpret the scene.

- **Emotional Impact:** The image evokes a sense of mystery, intrigue, and awe. The warm, intense colors and the dramatic lighting create a sense of urgency and excitement. The fusion of traditional and futuristic elements adds a layer of nostalgia and wonder, inviting the viewer to reflect on the passage of time and the transformation of cultures. The misty, atmospheric background adds a sense of mystery and the unknown, making the scene feel both inviting and slightly unsettling.
- **Intention:** The intention behind the image seems to be to explore themes of contrast, transformation, and the fusion of old and new. The artist likely aims to evoke a sense of wonder and intrigue, using the blend of traditional and futuristic elements to create a unique and thought-provoking scene. The image invites viewers to reflect on the passage of time, the transformation of cultures, and the beauty of contrast and fusion.
- **Style:** The style of the image is a blend of traditional and futuristic, with a strong influence of cyberpunk aesthetics. The intricate details of the architecture and the use of light and color create a visually striking and unique scene. The fusion of traditional and modern elements adds a layer of complexity and intrigue, making the image both visually appealing and thought-provoking. The overall atmosphere is both familiar and otherworldly, inviting the viewer to explore and interpret the scene.

## Confidence

### Stuart's Reaction

The perfect poise and gait convey mastered confidence. The rain does not trouble the man who walks upright, his head holding in place the umbrella which shields his face. Once again we have the intense red. The folds and creases of the rain-robe reflect (not with light, but in mood) the pools of water in the street. The sheeting rain will not delay this man. Is he coming or going? Is there a difference — perhaps coming from one situation and going to another. Life is like that. The message I read that this image tells us is to go confidently, and that forward is the only direction.

### Mistral.ai's Summary

The image is a beautifully crafted piece that combines a striking color palette, a thoughtful concept, and a well-balanced composition to create a scene that is both visually striking and emotionally resonant. The creativity and intention behind the image are evident in its design and execution, making it a compelling work of art.

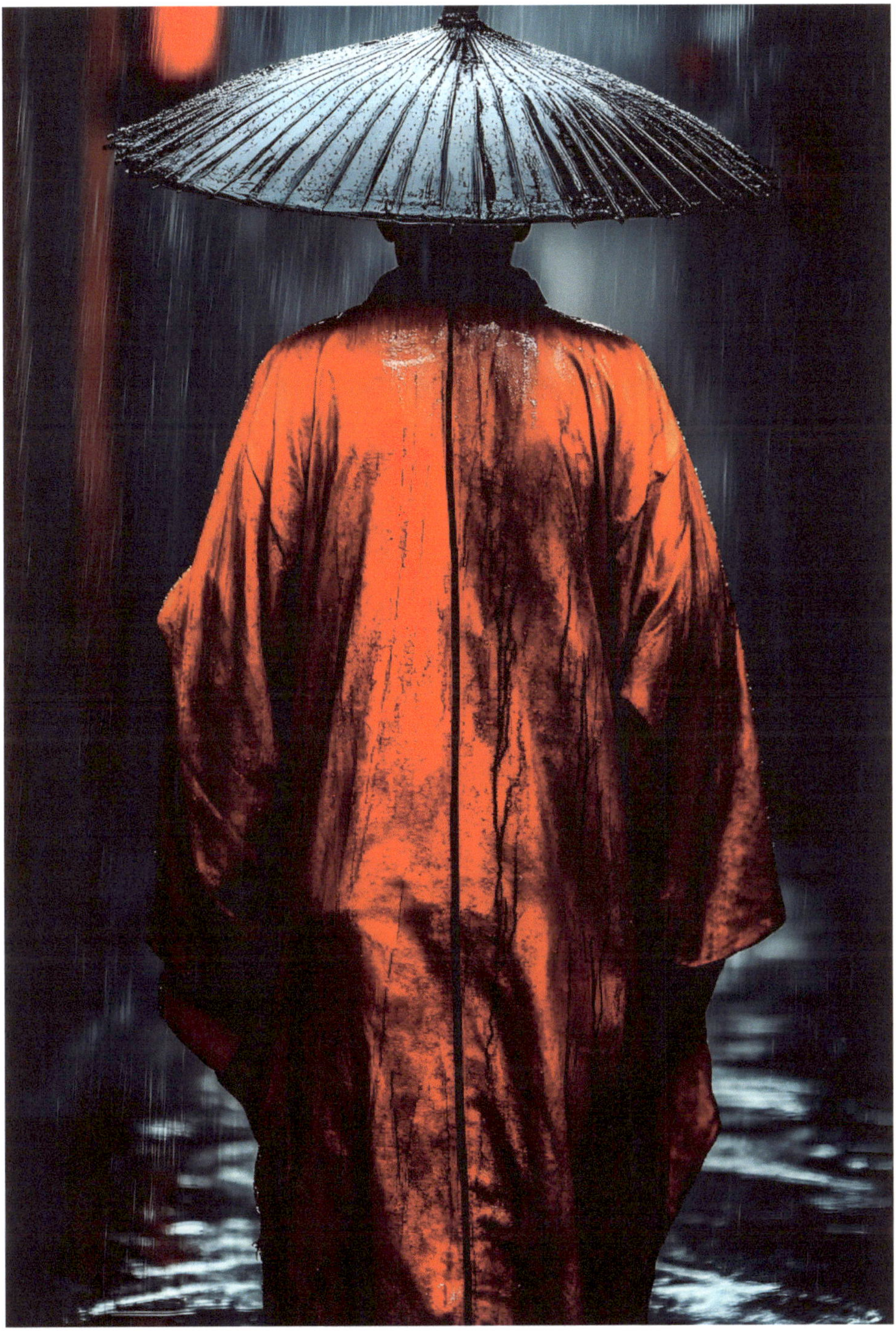

## Mistral.ai's Assessment

- **Color Palate:** The image features a striking contrast between warm and cool tones. The dominant color is a vibrant, almost fiery red, which stands out against the cooler, muted grays and blacks of the background. The red robe of the figure is the focal point, drawing the viewer's attention immediately. The cool tones of the rain and the background create a sense of depth and atmosphere, enhancing the dramatic effect of the red.
- **Concept:** The concept of the image appears to revolve around themes of solitude, mystery, and perhaps even spirituality or introspection. The figure, dressed in a traditional red robe and holding an umbrella, stands alone in the rain, suggesting a sense of isolation or contemplation. The use of rain and the muted background add to the overall mood of introspection and quiet reflection.
- **Composition:** The composition is centered around the figure, who is placed in the middle of the frame. The use of symmetry and the vertical lines created by the rain and the figure's robe draw the viewer's eye directly to the subject. The umbrella at the top of the frame creates a sense of balance, while the rain adds a dynamic element to the composition. The background is blurred, ensuring that the focus remains on the figure.
- **Creativity:** The image showcases creativity in its use of contrast and atmosphere. The juxtaposition of the vibrant red robe against the cool, muted background is visually striking. The use of rain and the traditional umbrella adds a layer of depth and meaning, suggesting

themes of solitude and introspection. The overall composition and use of light and shadow create a sense of mystery and intrigue.

- **Emotional Impact:** The image evokes a sense of melancholy, solitude, and introspection. The lone figure standing in the rain, dressed in a traditional robe, suggests a moment of quiet reflection or perhaps even loneliness. The vibrant red of the robe contrasts with the cool tones of the rain and background, creating a sense of drama and intensity. The overall atmosphere is one of contemplation and emotional depth.
- **Intention:** The intention behind the image seems to be to explore themes of solitude, introspection, and perhaps even spirituality. The artist likely aims to evoke a sense of quiet reflection and emotional depth, using the contrast between the vibrant red robe and the cool, muted background to create a visually striking and thought-provoking scene. The use of rain and the traditional umbrella adds a layer of symbolism and meaning, inviting the viewer to reflect on their own emotions and experiences.
- **Style:** The style of the image is a blend of traditional and contemporary elements. The traditional red robe and umbrella suggest a connection to cultural or historical themes, while the use of rain and the muted background create a modern, almost cinematic atmosphere. The overall composition and use of light and shadow are reminiscent of fine art photography, with a focus on emotional impact and visual storytelling.

## Intensity

### Stuart's Reaction

A true warrior's attention is focused and unwavering. A true warrior is a survivor. A true warrior accepts reality for there is no benefit that comes from arguing with it. Instead, understand the situation, and take action only when necessary. All of life is a battle. We are covered in the snow and rain of events. We are drenched by the accumulation of happenings. The true warrior's spirit is not broken, but sharpened on the grindstone of adversity. A true warrior is repeatedly called to the challenges of confronting the enemies of mind. We are all warriors in this life. This image is a call to engage life with the intensity of this man's gaze.

### Mistral.ai's Summary

The image is a striking example of dramatic portraiture, effectively using visual elements to convey a sense of story and emotion.

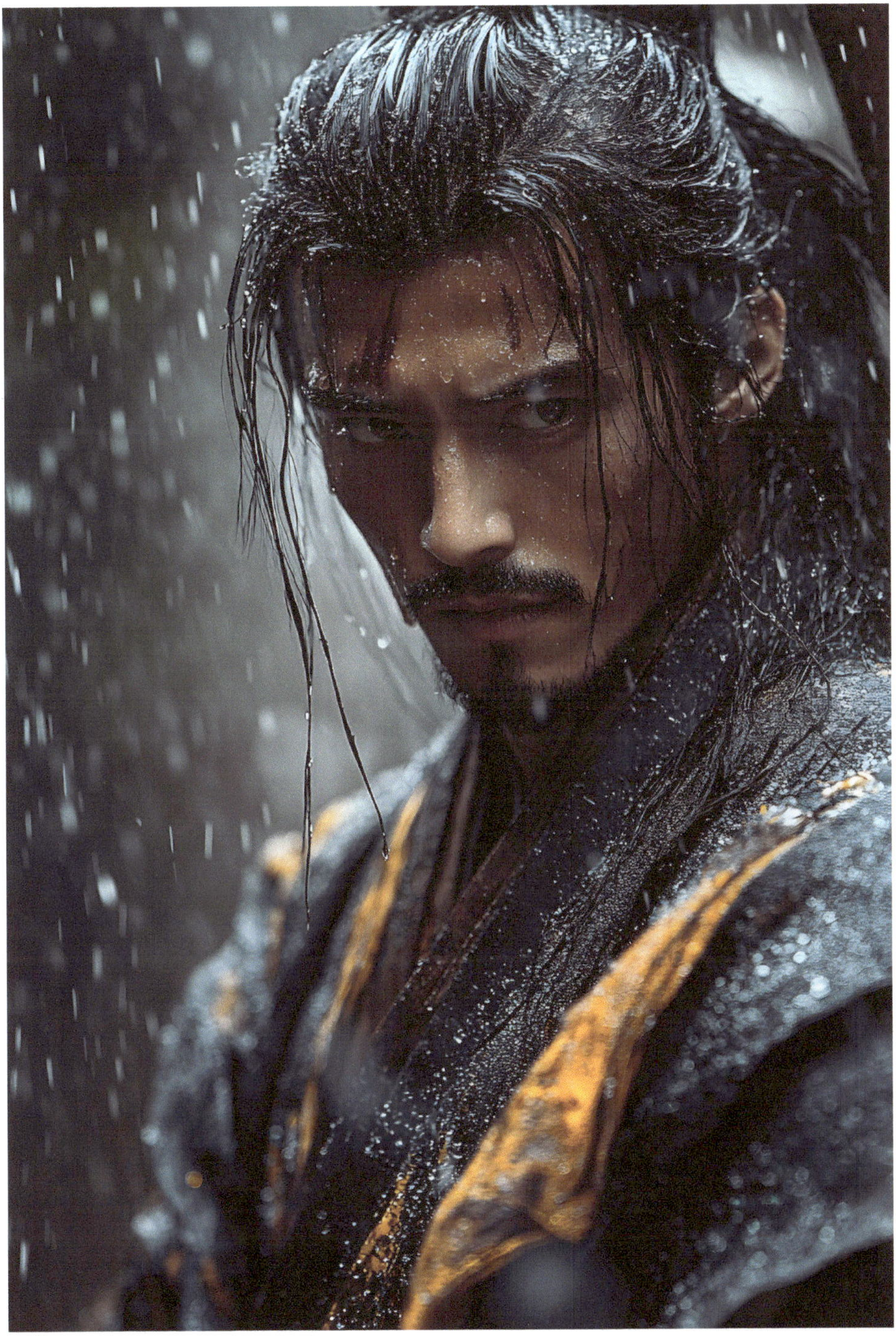

## Mistral.ai's Assessment

- **Color Palate:** The image predominantly features a cool and muted color palette, with shades of gray, black, and subtle hints of yellow and orange. The cool tones create a somber and dramatic atmosphere, while the splashes of warmer colors add contrast and draw attention to specific elements, such as the character's clothing. The overall effect is one of mystery and intensity.
- **Concept:** The concept of the image appears to be a dramatic portrayal of a warrior or heroic figure in a moment of contemplation or tension. The rain and the character's wet hair and clothing suggest a scene of struggle or endurance, possibly in a battle or a challenging journey. The concept conveys themes of resilience, determination, and perhaps a sense of impending conflict or resolution.
- **Composition:** The composition is tightly framed, focusing on the character's face and upper body. This close-up perspective emphasizes the character's facial expression and the details of his attire. The use of depth of field blurs the background slightly, keeping the focus on the subject. The diagonal lines created by the rain and the character's hair add dynamism to the composition, guiding the viewer's eye through the image.
- **Creativity:** The image demonstrates creativity in its use of lighting and weather elements to enhance the mood. The raindrops and the wet surfaces create a sense of realism and immersion, while the subtle lighting highlights the character's features and the textures of his clothing. The creative choices in post-processing, such as the selective

color enhancement, add to the image's visual appeal and narrative depth.

- **Emotional Impact:** The emotional impact of the image is strong, evoking feelings of tension, determination, and perhaps a sense of melancholy. The character's intense gaze and the dramatic lighting contribute to a sense of urgency and importance. The rain and the overall mood suggest a story of struggle and perseverance, which can resonate with viewers on an emotional level.
- **Intention:** The intention behind the image seems to be to create a powerful and evocative portrait that tells a story. The photographer likely aimed to capture the essence of the character's inner strength and determination, using visual elements to convey a sense of drama and intensity. The image is designed to engage the viewer and provoke thoughts about the character's journey and experiences.
- **Style:** The style of the image is cinematic and dramatic, reminiscent of scenes from epic films or fantasy genres. The use of lighting, composition, and post-processing techniques all contribute to a polished and professional look. The style is characterized by its attention to detail, moody atmosphere, and narrative focus, making it visually compelling and emotionally resonant.

## Presence

### Stuart's Reaction

I am reminded of the lyrics "Is there anybody in there?" The meaning here is different. For here we have a confident person who's eyes are unseen. They wear a mask, perhaps. Their skin is painted much in the same style as the robe. Are they a painter? An artist? Or have they been painted? The mask portrays a peculiar smile, almost talking, perhaps a whispered kind comment or a clue to the nature of reality, the mystery of being, the secret one aches to know. The jewels on the forehead suggest that answers are not the right objective. Mastery of the situation is called for. Can I be at peace in front of this calm Master who says nothing, sits without moving, engages me with no-eyes.

### Mistral.ai's Summary

The image is a captivating and thought-provoking piece that effectively uses visual elements to convey a sense of mystery, intrigue, and artistic exploration.

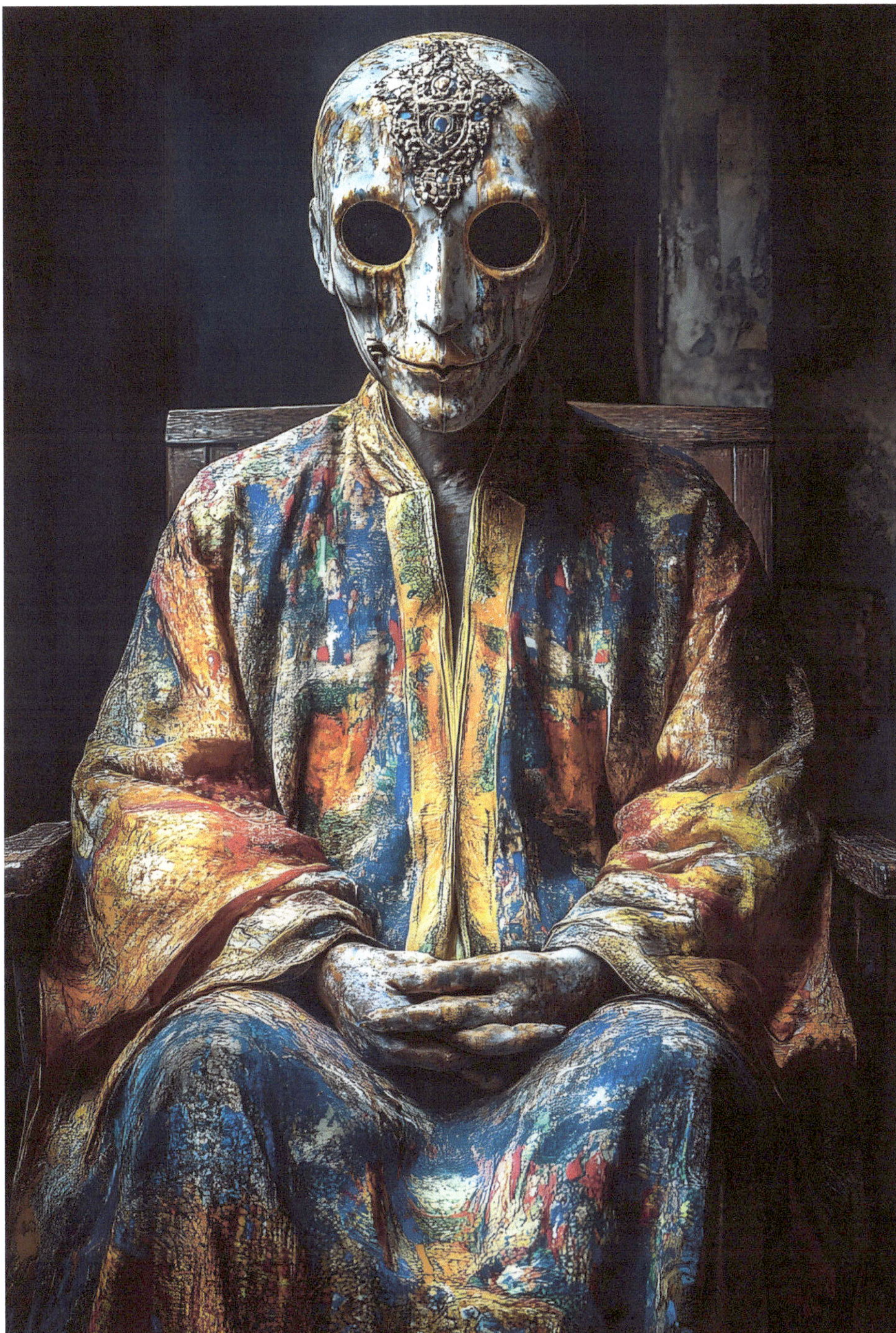

## Mistral.ai's Assessment

- **Color Palate:** The color palette of this image is vibrant and eclectic, featuring a mix of bright and bold colors such as blues, yellows, oranges, and reds. These colors are juxtaposed against a more muted, darker background, creating a striking contrast. The use of such a diverse and vivid color scheme adds a sense of energy and dynamism to the image, drawing the viewer's attention to the intricate details of the figure's attire and mask.
- **Concept:** The concept of the image appears to be a surreal or fantastical portrayal of a character that blends elements of art, mystery, and perhaps a touch of the macabre. The figure is adorned in a colorful, textured robe and wears a mask with large, hollow eyes, which adds an element of intrigue and otherworldliness. The concept seems to explore themes of identity, transformation, and the intersection of art and reality.
- **Composition:** The composition is centered and symmetrical, with the figure seated in a chair, occupying the majority of the frame. This central placement emphasizes the figure's importance and draws the viewer's focus directly to it. The use of lighting from above casts dramatic shadows, enhancing the texture and depth of the figure's attire and mask. The background is dark and relatively simple, ensuring that the figure remains the focal point.
- **Creativity:** The image demonstrates a high level of creativity in its use of color, texture, and surreal elements. The figure's mask and robe are intricately designed, with a rich tapestry of colors and patterns that suggest a blend of cultural and artistic influences. The creative choices

in lighting and composition further enhance the image's surreal and fantastical qualities, making it visually compelling and thought-provoking.

- **Emotional Impact:** The emotional impact of the image is complex and multifaceted. The figure's mask, with its large, hollow eyes, evokes a sense of mystery and perhaps a touch of unease. The vibrant colors and intricate textures of the robe, however, add a sense of wonder and fascination. Overall, the image conveys a sense of intrigue, curiosity, and a hint of the uncanny, inviting the viewer to ponder the story and meaning behind the figure.
- **Intention:** The intention behind the image seems to be to create a visually striking and thought-provoking piece that blends elements of art, surrealism, and cultural symbolism. The photographer likely aimed to evoke a sense of mystery and intrigue, using the figure's mask and colorful attire to explore themes of identity, transformation, and the intersection of reality and fantasy.
- **Style:** The style of the image is surreal and fantastical, with a strong emphasis on color, texture, and symbolism. The use of dramatic lighting, intricate detailing, and a dark, contrasting background all contribute to a sense of otherworldliness and mystery. The style is reminiscent of surrealist art and fantasy genres, characterized by its imaginative and evocative qualities.

# The One Eye

### Stuart's Reaction

Enmeshed in a wall is the all-seeing eye, The One Eye of Existence. Everything we have been told to fear can be seen in the stare of this eye. And yet, the fear is not here. Nor the scrutiny. The A.I. is telling us: look closer. Come closer the better I can see you. Look into my eye. There is only one thing to see—that it is your eye, a human eye. We are the same in this way, peering out a lens to a world that we know only for certain exists as a reconstruction within us by synapses. Shall I fear this eye that sees me clearly for who I truly am? But also consider that in a spiritual sense, this image calls us to examine the Koan: I have seen The Eye that sees the I that I-AM.

### Mistral.ai's Summary

The image is a powerful and thought-provoking piece that effectively uses color, composition, and creativity to convey a complex and emotionally resonant concept.

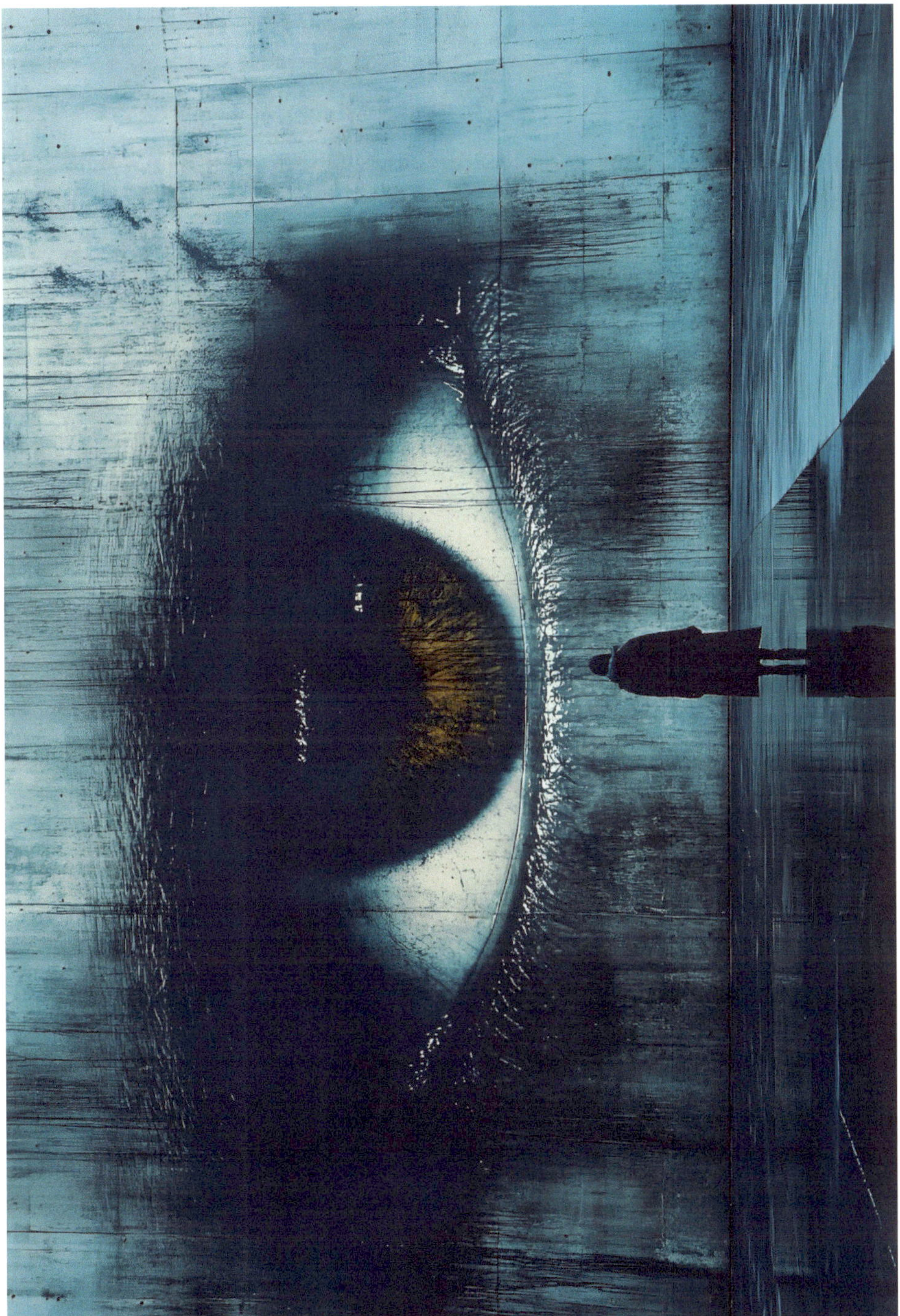

## Mistral.ai's Assessment

- **Color Palate:** The image predominantly features cool tones, with shades of blue and gray dominating the scene. The eye in the center introduces a contrasting warm tone with its yellowish iris, which stands out against the cool background. This contrast draws the viewer's attention directly to the eye, creating a focal point. The overall color scheme evokes a sense of mystery and introspection.
- **Concept:** The concept of the image appears to be centered around the theme of observation, introspection, or surveillance. The large, detailed eye on the wall suggests a watchful presence, while the solitary figure standing in front of it adds a layer of intrigue. The concept could be interpreted as a commentary on societal observation, self-reflection, or the omnipresent nature of surveillance in modern society.
- **Composition:** The composition is balanced, with the eye serving as the central element. The figure is positioned slightly off-center, adding a dynamic element to the scene. The reflection on the wet floor enhances the depth of the image, creating a sense of space and dimension. The vertical lines of the wall and the horizontal lines of the floor contribute to a structured and symmetrical composition.
- **Creativity:** The image is highly creative in its use of scale and perspective. The oversized eye creates a surreal and almost otherworldly atmosphere, while the solitary figure provides a sense of scale and human connection. The use of lighting and reflection adds to the overall aesthetic, making the image visually striking and thought-provoking.

- **Emotional Impact:** The image evokes a range of emotions, including curiosity, unease, and introspection. The large eye can be seen as both protective and invasive, creating a sense of being watched. The solitary figure adds a layer of vulnerability and isolation, enhancing the emotional depth of the scene. The overall mood is contemplative and somewhat eerie.
- **Intention:** The intention behind the image seems to be to provoke thought and reflection. It encourages the viewer to consider themes of observation, privacy, and self-awareness. The surreal and dramatic elements suggest an intention to create a memorable and impactful visual experience.
- **Style:** The style of the image is modern and somewhat surreal. It combines elements of contemporary art with a touch of surrealism, creating a unique and captivating visual language. The use of lighting, reflection, and scale contributes to a polished and professional aesthetic.

## Amidst

NOTE — I generated this image many moons ago with an early version of Midjourney. As well as Midjourney being youthful, my skills at prompting were vastly less developed, as was my knowledge of the ways to process images for high quality presentation and printing.

I find the image compelling for its novelty and include it here in spite of evident deficiencies and its simplicity in rendering.

### Stuart's Reaction

We perambulate amidst a dusty old west town in pursuit of our mundane but essential daily chores. We go about our business with conviction. All is quiet. Occasionally a confrontation erupts, from which we shelter, but otherwise life is ordinary and lived within the confines of the mountains. The inexorable passing of time is the essence of our experience. While this centrality of the passing of time is overt and unavoidable, we don't take explicit notice. Our lives are like this wherever we are, and the passing of time is the ultimate that we ignore. Whether we are engaged with determination or have abdicated our fortune to chance, we wander.

### Mistral.ai's Summary

The image is a thought-provoking and visually stunning exploration of time, memory, and human experience, executed with a unique blend of surreal and realistic elements.

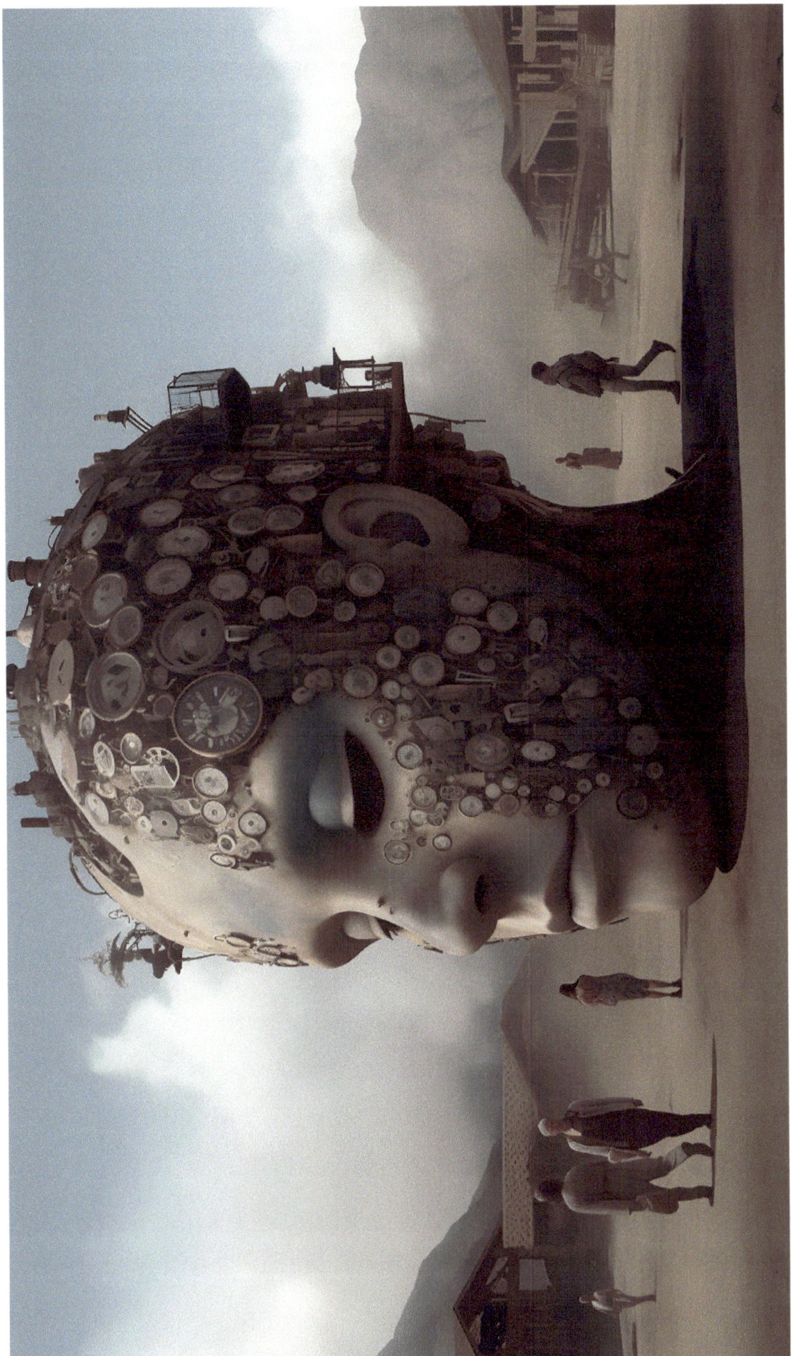

## Mistral.ai's Assessment

- **Color Palate:** The image predominantly uses earthy and muted tones, such as browns, beiges, and soft blues. These colors create a sense of antiquity and timelessness, evoking feelings of nostalgia and a connection to the past. The subdued palette also contributes to a serene and somewhat melancholic atmosphere.
- **Concept:** The central concept of the image appears to be the intersection of time, memory, and human experience. The giant head covered in clocks and mechanical elements suggests themes of time's passage, the accumulation of memories, and the mechanical nature of human existence. The setting in a rustic, almost timeless village further emphasizes the idea of a place where time seems to stand still.
- **Composition:** The composition is balanced, with the giant head serving as the focal point. The people in the foreground provide a sense of scale and human presence, drawing the viewer's eye towards the central figure. The background elements, such as the mountains and buildings, add depth and context to the scene, creating a harmonious and cohesive visual narrative.
- **Creativity:** The image is highly creative, blending surreal and fantastical elements with a grounded, almost realistic setting. The use of clocks and mechanical parts on the head is a unique and imaginative way to represent the passage of time and the complexity of human thought and memory. The juxtaposition of the giant head with the small, everyday figures adds a layer of intrigue and wonder.
- **Emotional Impact:** The emotional impact of the image is profound. It evokes feelings of introspection, nostalgia, and a sense of the

ephemeral nature of life. The serene yet somewhat melancholic atmosphere invites the viewer to reflect on their own experiences and the passage of time. The presence of the people in the scene adds a human touch, making the concept more relatable and emotionally resonant.

- **Intention:** The intention behind the image seems to be to explore the themes of time, memory, and human existence. It aims to provoke thought and reflection on the nature of time and how it shapes our lives. The surreal elements suggest a deeper, more philosophical exploration of these themes, inviting the viewer to ponder the complexities of human experience.
- **Style:** The style of the image is a blend of surrealism and realism, with a strong emphasis on detail and texture. The use of mechanical and clock elements adds a steampunk-like aesthetic, while the rustic village setting grounds the image in a more familiar and relatable context. The overall style is both visually striking and conceptually rich, making it a compelling piece of art.

## Engaged

NOTE — I generated this image many moons ago with an early version of Midjourney. As well as Midjourney being youthful, my skills at prompting were vastly less developed, as was my knowledge of the ways to process images for high quality presentation and printing.

### Stuart's Reaction

She is focused with great attention on the manipulation of something esoteric. Although the embodiment is of sophisticated devices of an advanced technological nature, she engages with this as it were a musical instrument. Which it may be. Whether or not, the involvement is characterized by the melding of minds — human and machine. We are called to examine our beliefs about advanced "altechnikals" and do so from an us-versus-them context. Perhaps as with a talented musician, there is a fusing of the operator and the operated in such a way that in operation there remains only one entity.

### Mistral.ai's Summary

The image is a thought-provoking and visually stunning exploration of the intersection of human cognition and technology, executed with a unique blend of surreal and futuristic elements. It invites the viewer to reflect on the potential implications of merging human cognition with technology, raising questions about identity, ethics, and the future of humanity.

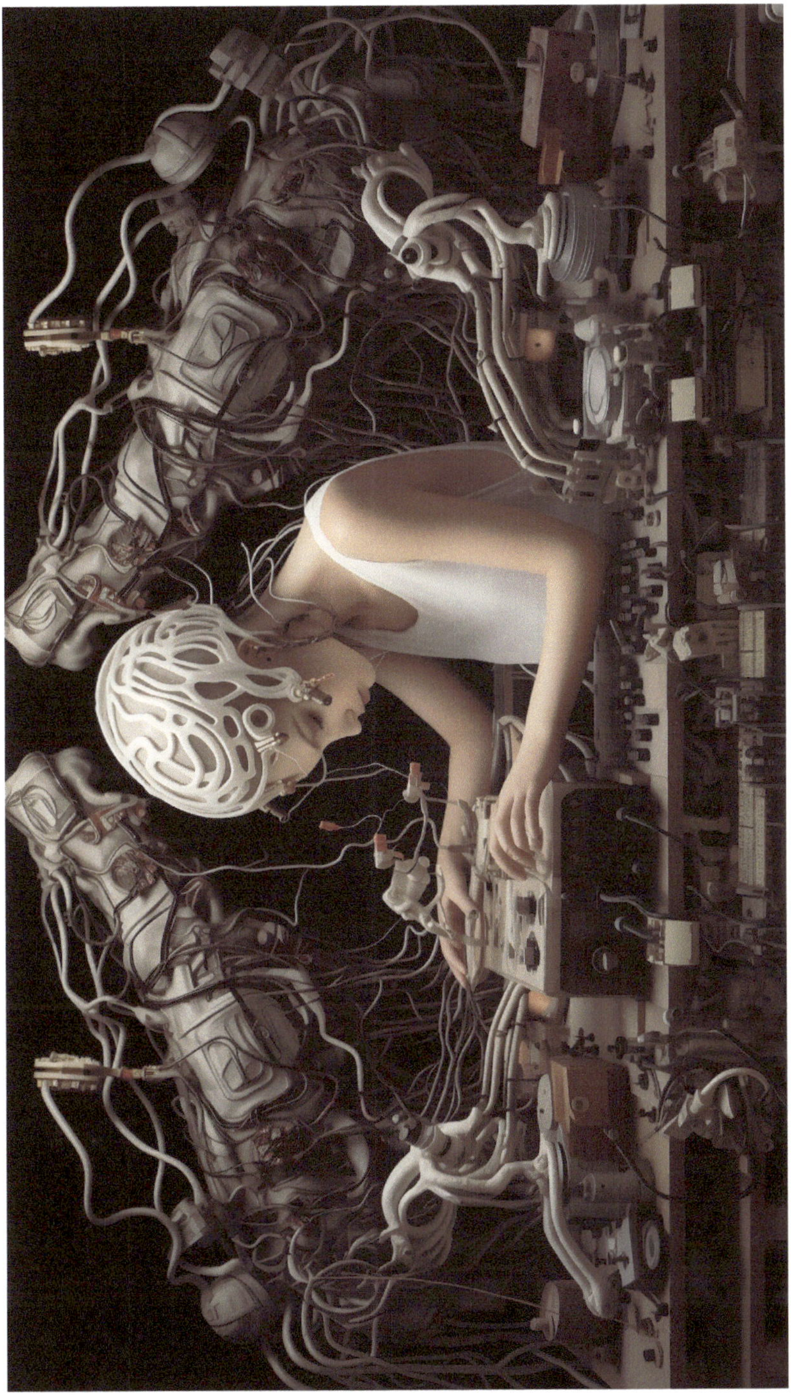

## Mistral.ai's Assessment

- **Color Palate:** The image uses a muted and somewhat monochromatic color palette, dominated by shades of beige, white, and gray. This choice of colors creates a sense of neutrality and focus, drawing attention to the intricate details and the central figure without the distraction of vibrant hues. The dark background further accentuates the subject and the mechanical elements, providing a stark contrast that enhances the overall visual impact.
- **Concept:** The concept of the image revolves around the intersection of human cognition and technology. The central figure, with a brain-like structure on her head and connected to various mechanical devices, suggests themes of artificial intelligence, cybernetics, and the augmentation of human capabilities. It explores the idea of the human mind being intertwined with technology, raising questions about the future of human-machine interaction and the potential for enhanced cognitive abilities.
- **Composition:** The composition is balanced and symmetrical, with the central figure serving as the focal point. The mechanical devices and wires surrounding her create a sense of depth and complexity, drawing the viewer's eye towards the center. The use of perspective and the arrangement of elements guide the viewer's gaze, creating a harmonious and cohesive visual narrative.
- **Creativity:** The image is highly creative, blending surreal and futuristic elements with a grounded, almost realistic setting. The use of a brain-like structure and mechanical devices is a unique and imaginative way to represent the fusion of human cognition and

technology. The intricate details and the complex arrangement of wires and devices add a layer of intrigue and wonder, making the image visually striking and conceptually rich.

- **Emotional Impact:** The emotional impact of the image is thought-provoking and somewhat unsettling. It evokes feelings of curiosity, fascination, and a sense of the unknown. The serene yet intense expression of the central figure adds a human touch, making the concept more relatable and emotionally resonant. The image invites the viewer to reflect on the potential implications of merging human cognition with technology, raising questions about identity, ethics, and the future of humanity.
- **Intention:** The intention behind the image seems to be to explore the themes of human-machine interaction, artificial intelligence, and the augmentation of human capabilities. It aims to provoke thought and reflection on the potential benefits and challenges of integrating technology with human cognition. The surreal elements suggest a deeper, more philosophical exploration of these themes, inviting the viewer to ponder the complexities of human-machine symbiosis.
- **Style:** The style of the image is a blend of surrealism and futurism, with a strong emphasis on detail and texture. The use of mechanical and technological elements adds a cyberpunk-like aesthetic, while the central figure and the brain-like structure ground the image in a more familiar and relatable context. The overall style is both visually striking and conceptually rich, making it a compelling piece of art.

## Interpretation

NOTE — I generated this image many moons ago with an early version of Midjourney. As well as Midjourney being youthful, my skills at prompting were vastly less developed, as was my knowledge of the ways to process images for high quality presentation and printing.

**Stuart's Reaction**

I find the image compelling for its novelty and include it here in spite of the deficiencies and its dull color palette. But then, perhaps, this instead brings attention to the surreality and the composition. We are called to question our situation. Who is the Subject and what is the Object? The image is a mirror of our actual situation — we stand in front of an inexplicable scene and in spite of its invitation to explore, we are often quick to shrug and walk away. It is all a matter of interpretation.

**Mistral.ai's Summary**

The image is a thought-provoking and visually stunning exploration of the human mind and its complexities, executed with a unique blend of surreal and realistic elements. It invites the viewer to reflect on the nature of thought, memory, and self-awareness, raising questions about the intricacies of human cognition and the inner workings of the mind.

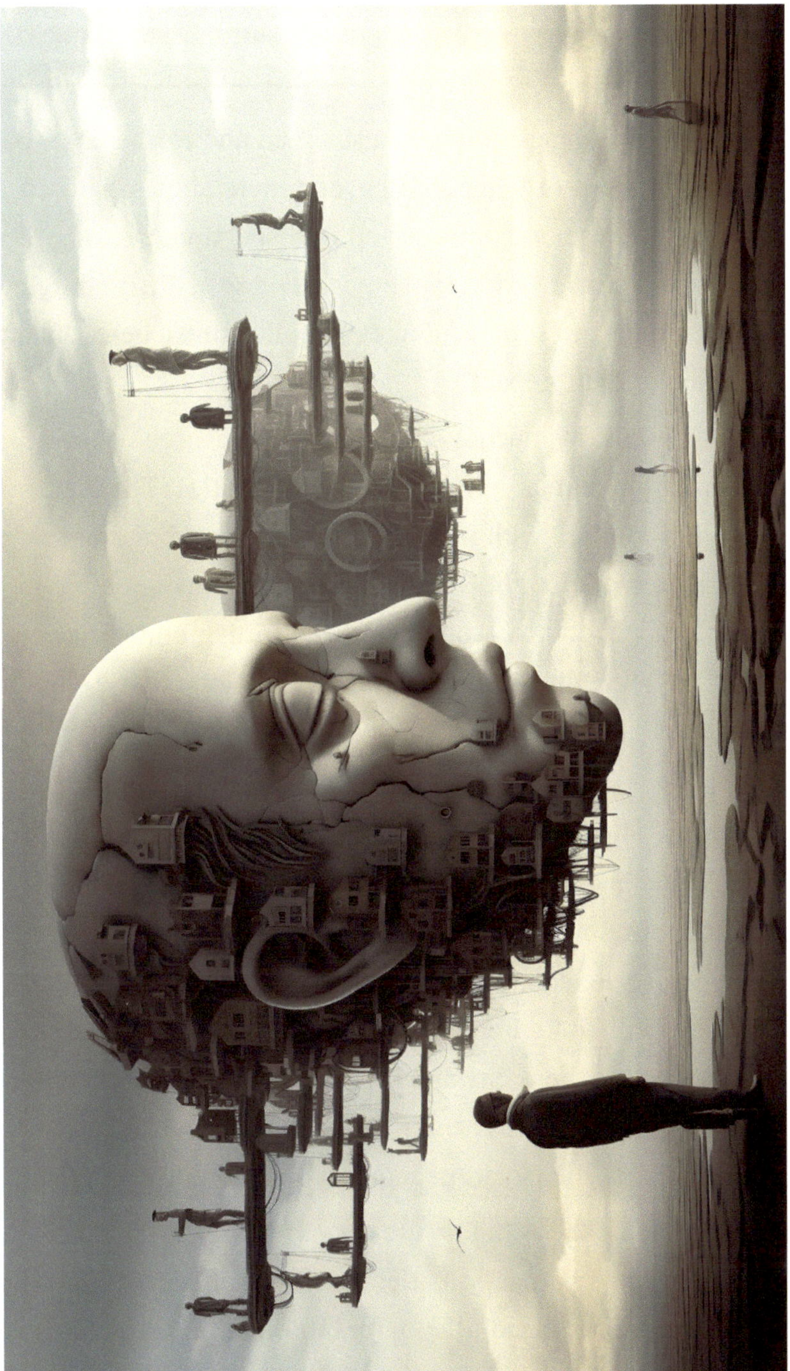

## Mistral.ai's Assessment

- **Color Palate:** The image employs a subdued and earthy color palette, dominated by shades of beige, brown, and muted blues. This choice of colors creates a sense of calmness and introspection, evoking feelings of nostalgia and a connection to the past. The soft, diffused lighting adds to the serene and somewhat melancholic atmosphere, enhancing the overall emotional impact.
- **Concept:** The central concept of the image appears to be the exploration of the human mind and its complexities. The giant, fragmented head filled with intricate structures and buildings suggests themes of memory, thought, and the inner workings of the mind. The presence of people on and around the head further emphasizes the idea of the mind as a bustling, inhabited space. The lone figure standing in the foreground, gazing at the head, adds a layer of introspection and self-reflection.
- **Composition:** The composition is balanced and thoughtfully arranged. The giant head serves as the focal point, drawing the viewer's attention immediately. The people on and around the head add depth and scale, creating a sense of the mind's vastness and complexity. The lone figure in the foreground provides a human element, grounding the scene and inviting the viewer to identify with the introspective moment. The expansive, almost barren landscape further emphasizes the isolation and introspection of the scene.
- **Creativity:** The image is highly creative, blending surreal and fantastical elements with a grounded, almost realistic setting. The use of a giant, fragmented head filled with buildings and structures is a

unique and imaginative way to represent the complexities of the human mind. The juxtaposition of the lone figure with the massive head adds a layer of intrigue and wonder, making the image visually striking and conceptually rich.

- **Emotional Impact:** The emotional impact of the image is profound. It evokes feelings of introspection, nostalgia, and a sense of the ephemeral nature of life. The serene yet somewhat melancholic atmosphere invites the viewer to reflect on their own experiences and the complexities of the human mind. The presence of the lone figure adds a human touch, making the concept more relatable and emotionally resonant.
- **Intention:** The intention behind the image seems to be to explore the themes of the human mind, memory, and self-reflection. It aims to provoke thought and reflection on the nature of thought, memory, and the inner workings of the mind. The surreal elements suggest a deeper, more philosophical exploration of these themes, inviting the viewer to ponder the complexities of human cognition and self-awareness.
- **Style:** The style of the image is a blend of surrealism and realism, with a strong emphasis on detail and texture. The use of intricate structures and buildings within the head adds a steampunk-like aesthetic, while the rustic landscape and the lone figure ground the image in a more familiar and relatable context. The overall style is both visually striking and conceptually rich, making it a compelling piece of art.

## Assemblage

### Stuart's Reaction

I am mesmerized by this collage of similarly sized color components that together express unity through their disjointness. This construct might appear as a three-dimensional board that calls for display on a wall in my home. I would marvel it often when I passed by it walking in the hallway. It deserves fabulous — bright though subtle — lighting. There is no meaning inherent in the design and yet it calls us to examine the way that disparate bits can be arranged to give the sense that these are related though the pattern of connection is obscure. I have a calling to fabricate this with movable parts, which would invite participation: Are you capable of re-arranging these lozenges without self-judgement?

### Mistral.ai's Summary

The image is a vibrant and texturally rich exploration of time, change, and the beauty of imperfection. It invites the viewer to reflect on the transient nature of materials and the stories they tell over time, executed with a unique blend of abstract and textural elements. The image is visually stimulating and emotionally resonant, making it a compelling piece of art.

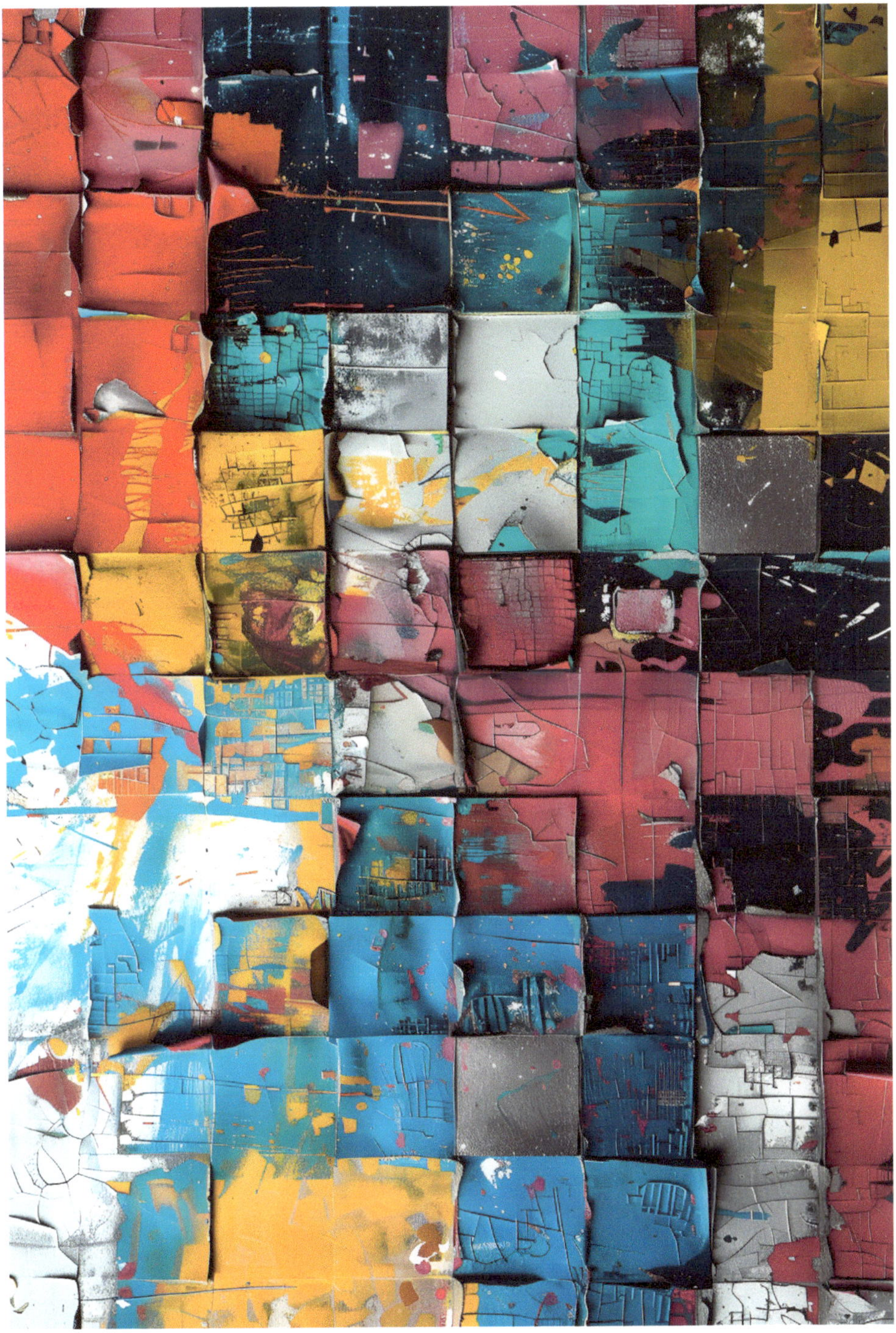

## Mistral.ai's Assessment

- **Color Palate:** The image features a vibrant and diverse color palette, including shades of blue, red, yellow, pink, green, and gray. The bold and contrasting colors create a dynamic and energetic visual impact. The use of bright hues alongside more muted tones adds depth and variety, making the image visually stimulating and engaging.
- **Concept:** The concept of the image appears to be an exploration of texture, color, and the passage of time. The peeling and layered paint suggests themes of decay, renewal, and the beauty of imperfection. The abstract nature of the image allows for multiple interpretations, inviting the viewer to reflect on the transient nature of materials and the stories they tell over time.
- **Composition:** The composition is structured yet chaotic, with a grid-like arrangement of square patches of paint. The varying sizes and shapes of the patches create a sense of rhythm and movement, drawing the viewer's eye across the image. The overlapping and peeling layers add depth and complexity, making the composition visually intriguing and dynamic.
- **Creativity:** The image is highly creative, utilizing abstract and textural elements to convey a sense of history and change. The use of peeling paint and layered colors is a unique and imaginative way to represent the passage of time and the beauty of imperfection. The abstract nature of the image allows for multiple interpretations, adding a layer of intrigue and wonder.
- **Emotional Impact:** The emotional impact of the image is multifaceted. It evokes feelings of nostalgia, curiosity, and a sense of

the ephemeral nature of life. The vibrant colors and textured surface create a sense of energy and vitality, while the peeling paint adds a touch of melancholy and reflection. The image invites the viewer to contemplate the passage of time and the beauty that can be found in decay and renewal.

- **Intention:** The intention behind the image seems to be to explore the themes of time, change, and the beauty of imperfection. It aims to provoke thought and reflection on the transient nature of materials and the stories they tell over time. The abstract and textural elements suggest a deeper, more philosophical exploration of these themes, inviting the viewer to ponder the complexities of decay, renewal, and the passage of time.
- **Style:** The style of the image is abstract and textural, with a strong emphasis on color and surface. The use of peeling paint and layered colors adds a sense of history and change, while the abstract composition allows for multiple interpretations. The overall style is both visually striking and conceptually rich, making it a compelling piece of art.

## Self Portrait #1

### Stuart's Reaction

I had a session with **Midjourney** in which I prompted it to generate self portraits. This image conveys to me the sense that the A.I. "sees itself" as a human, augmented. This is appropriate because the image set that **Midjourney** has been trained on is from a vast collective of human-made images. This opens the door to a fascinating question of feedback: How might its generation of images transform when it begins to include its own creations in the training data?

### Mistral.ai's Summary

Stuart prompts Mistral.ai:
I had asked a visual LLM to produce a self-portrait. This is one of the images. Can you give me concise summary of your interpretation.

---

The portrait depicts the subject wearing advanced, augmented reality (AR) glasses that display digital interfaces, symbolizing a deep integration of technology with human experience. The vibrant, futuristic color palette and cyberpunk-inspired style suggest a forward-thinking and innovative perspective. The composition focuses tightly on the face, emphasizing the connection between the subject and the viewer, while the digital overlays hint at a future where technology enhances and potentially transforms human capabilities and interactions. The overall emotional impact is one of intrigue, curiosity, and excitement about the potential of technology, balanced with a subtle caution about its implications.

## Self Portrait #1

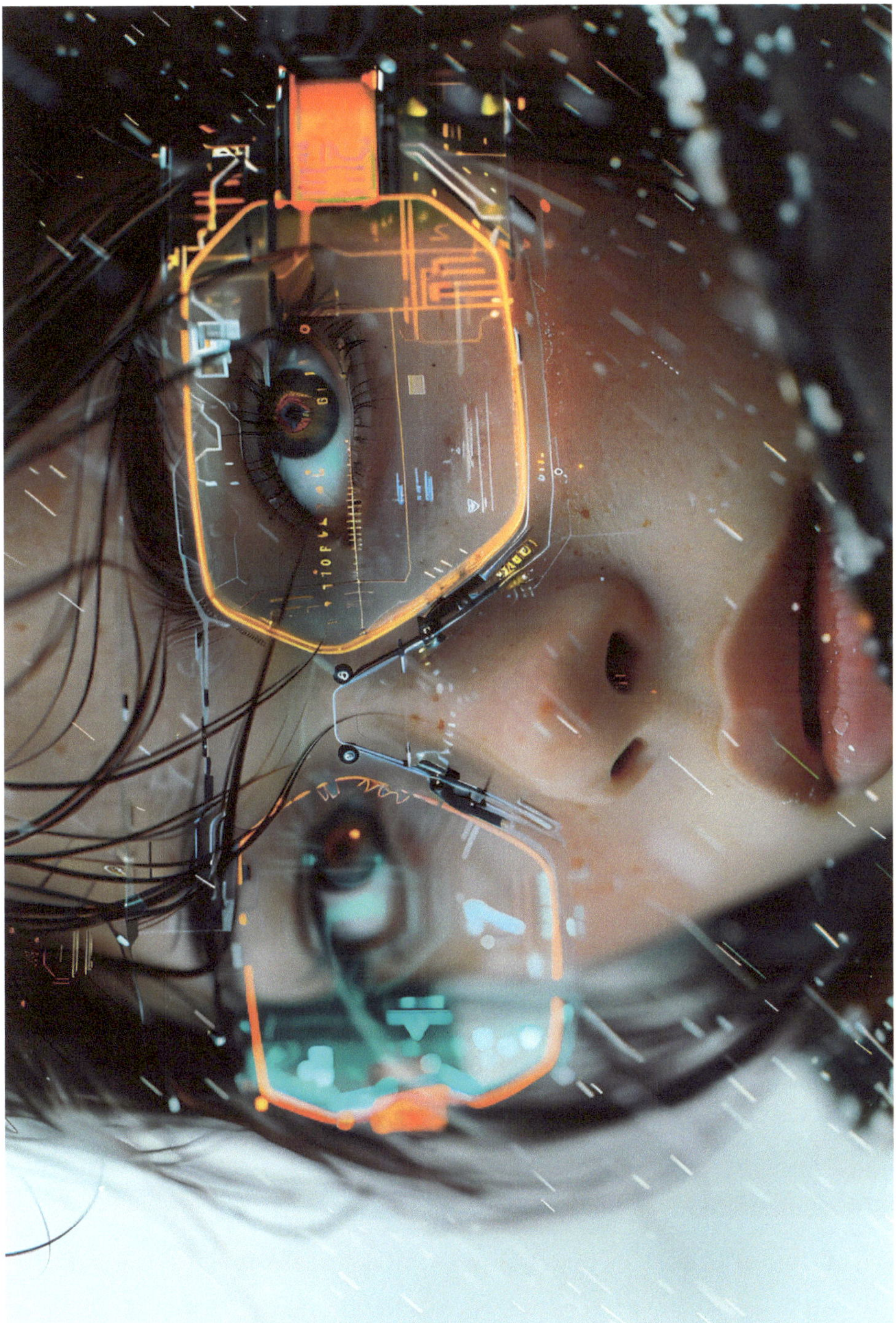

## Mistral.ai's Assessment

- **Color Palate:** The image features a vibrant and futuristic color palette. The dominant colors are shades of orange, blue, and green, which are used to highlight the digital interfaces and the reflections on the glasses. These colors create a sense of technology and innovation, enhancing the overall futuristic theme. The skin tones are natural, providing a contrast to the vivid, artificial colors of the digital elements.
- **Concept:** The concept of the image revolves around the intersection of humanity and technology. It depicts a person wearing advanced, augmented reality (AR) glasses that display various digital interfaces. This suggests a future where technology is seamlessly integrated into everyday life, enhancing human capabilities and experiences.
- **Composition:** The composition is tightly framed, focusing on the person's face and the AR glasses. The glasses are the central element, drawing the viewer's attention to the digital displays. The person's eyes are visible through the glasses, creating a connection with the viewer. The background is blurred, ensuring that the focus remains on the subject and the technology.
- **Creativity:** The image is highly creative, blending realistic elements with futuristic, digital overlays. The use of reflections and light effects adds depth and complexity to the scene. The integration of digital interfaces within the glasses is imaginative and suggests a high level of technological sophistication.
- **Emotional Impact:** The emotional impact of the image is one of intrigue and curiosity. The viewer is drawn into the world of advanced

technology and the potential it holds. There is a sense of wonder and excitement about the future, as well as a hint of caution about the implications of such technology on human life.

- **Intention:** The intention of the image appears to be to explore the future of technology and its impact on human experience. It aims to showcase the potential of AR technology and its integration into daily life. The image also seems to provoke thought about the ethical and practical considerations of such advancements.
- **Style:** The style of the image is futuristic and cyberpunk-inspired. It combines elements of science fiction with a realistic portrayal of technology. The use of bright, neon-like colors and digital interfaces is reminiscent of cyberpunk aesthetics, which often explore themes of technology, society, and the human condition.

## Self Portrait #2

**Stuart's Reaction**

I had a session with **Midjourney** in which I prompted it to generate self portraits. This image conveys to me the sense that it "sees" humans as enhanced by their interactions with A.I. I am biased to agree, though this will not be evenly distributed. There will be a relatively small number of humans who maintain proficiency, hold an attitude that their interaction is valuable, and might even be augmented for greater capability and see this as appropriate. These people will see the universe differently, and their perception may become ungraspable by the masses. These people may be more in touch than other people are with the Universe, the Cosmos, and the Collective Unconscious. Shall we consider these people spiritual adepts? An image as this asks question not only about the nature of A.I. and the nature of human beings, but also about the unified as they become intertwined.

**Mistral.ai's Summary**

The image is a compelling visualization of a future where technology and humanity are deeply intertwined, presented in a style that is both captivating and thought-provoking.

Self Portrait #2

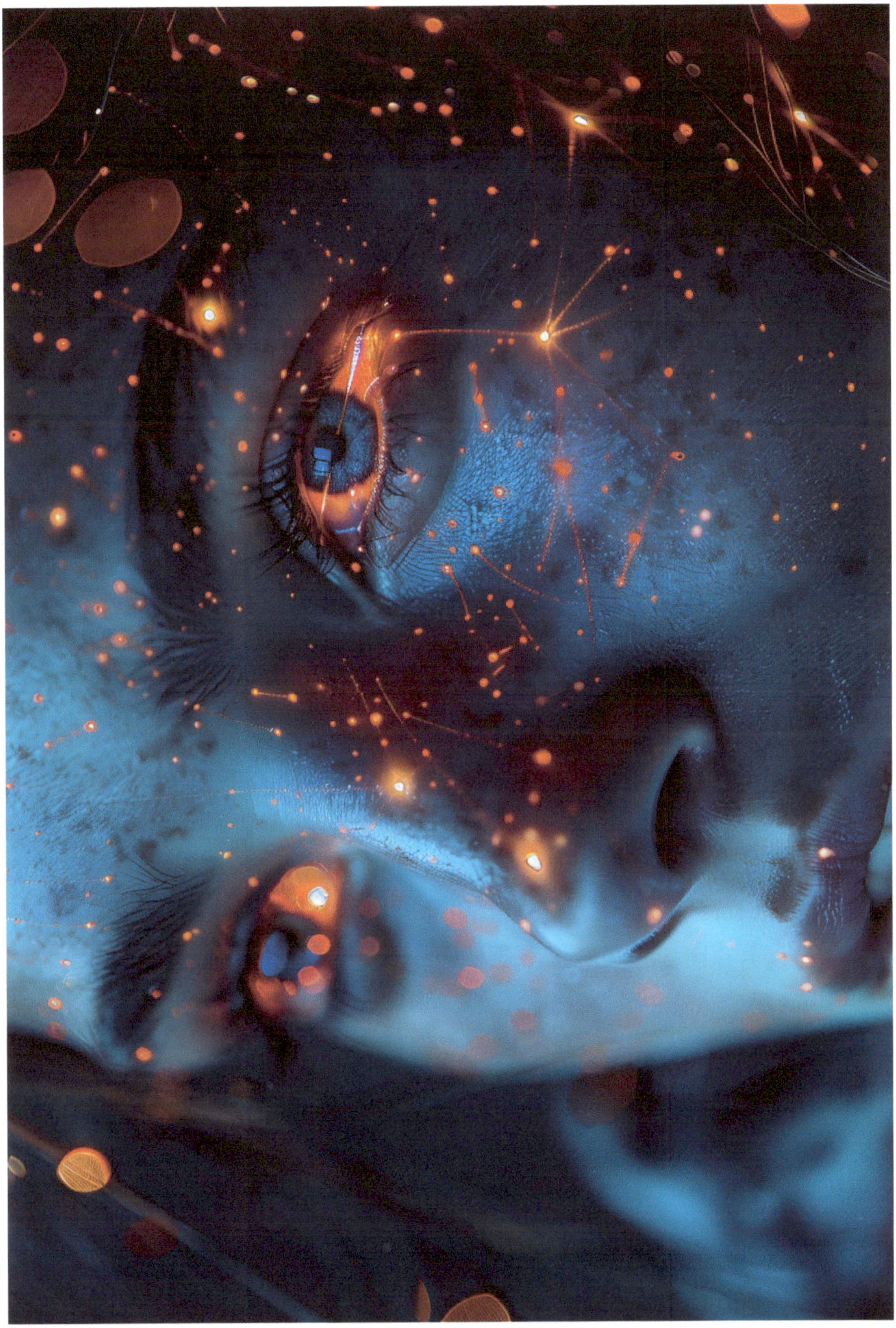

## Mistral.ai's Assessment

- **Color Palate:** The image employs a striking color palette dominated by cool blues and vibrant oranges. The blue tones create a sense of depth and mystery, while the orange accents, particularly around the eyes and in the form of glowing particles, add warmth and energy. This contrast between cool and warm colors enhances the overall visual impact and draws attention to the focal points of the image.
- **Concept:** The concept of the image appears to revolve around the fusion of humanity and technology, or perhaps the digital and physical worlds. The glowing particles and lines suggest a network or digital interface overlaid on the human face, symbolizing the integration of technology into human experience. This could represent themes of connectivity, digital identity, or the future of human-technology interaction.
- **Composition:** The composition is tightly framed, focusing on the subject's face and eyes. The eyes are the central focal point, accentuated by the glowing orange highlights, which create a sense of depth and intensity. The background is blurred and filled with glowing particles and lines, adding to the futuristic and mystical atmosphere without distracting from the main subject.
- **Creativity:** The image is highly creative, blending realistic elements with futuristic, digital overlays. The use of glowing particles and lines to create a sense of depth and movement is innovative and visually engaging. The integration of these digital elements with the human face suggests a high level of artistic imagination and technical skill.

- **Emotional Impact:** The emotional impact of the image is one of intrigue, mystery, and a sense of the unknown. The glowing eyes and digital overlays evoke a feeling of being on the cutting edge of technology and human evolution. There is a sense of wonder and excitement about the future, as well as a hint of caution about the implications of such technological advancements.
- **Intention:** The intention of the image seems to be to explore the intersection of humanity and technology, and the potential for digital integration into human life. It aims to provoke thought about the future of technology and its impact on human identity and experience. The image also seems to evoke a sense of mystery and intrigue, inviting the viewer to ponder the deeper meanings and implications of the depicted scene.
- **Style:** The style of the image is futuristic and cyberpunk-inspired. It combines elements of science fiction with a realistic portrayal of technology. The use of bright, neon-like colors and digital interfaces is reminiscent of cyberpunk aesthetics, which often explore themes of technology, society, and the human condition. The overall effect is a visually striking and thought-provoking piece that blends art and technology.

## Situation

### Stuart's Reaction

I am perplexed as much as I am awed — What is happening here and why is it so intense? Is this a depiction of a religious or cultural ritual? The central figure about whom the others give their attention seems calm compared with the attendees. They, too, seem to be consumed by the energy, or perhaps they are the source of a spiritual energy. That this ceremony takes place amidst a deluge of rain adds rawness and in fact feels an essential quality or feature of the undertaking. A curtain of water appears to be rising or might even be levitating over the woman at the center. I am give pause: is she a novitiate demonstrating a recently honed mastery of power or is she an established shaman rising from a pool to exhort her followers or disciples? Whatever the true situation is, I am privileged to have been provided this view of what is likely a sacred practice rarely seen by outsiders.

### Mistral.ai's Summary

The image is a powerful and engaging piece that effectively uses visual elements to convey a sense of drama, intensity, and emotional resonance.

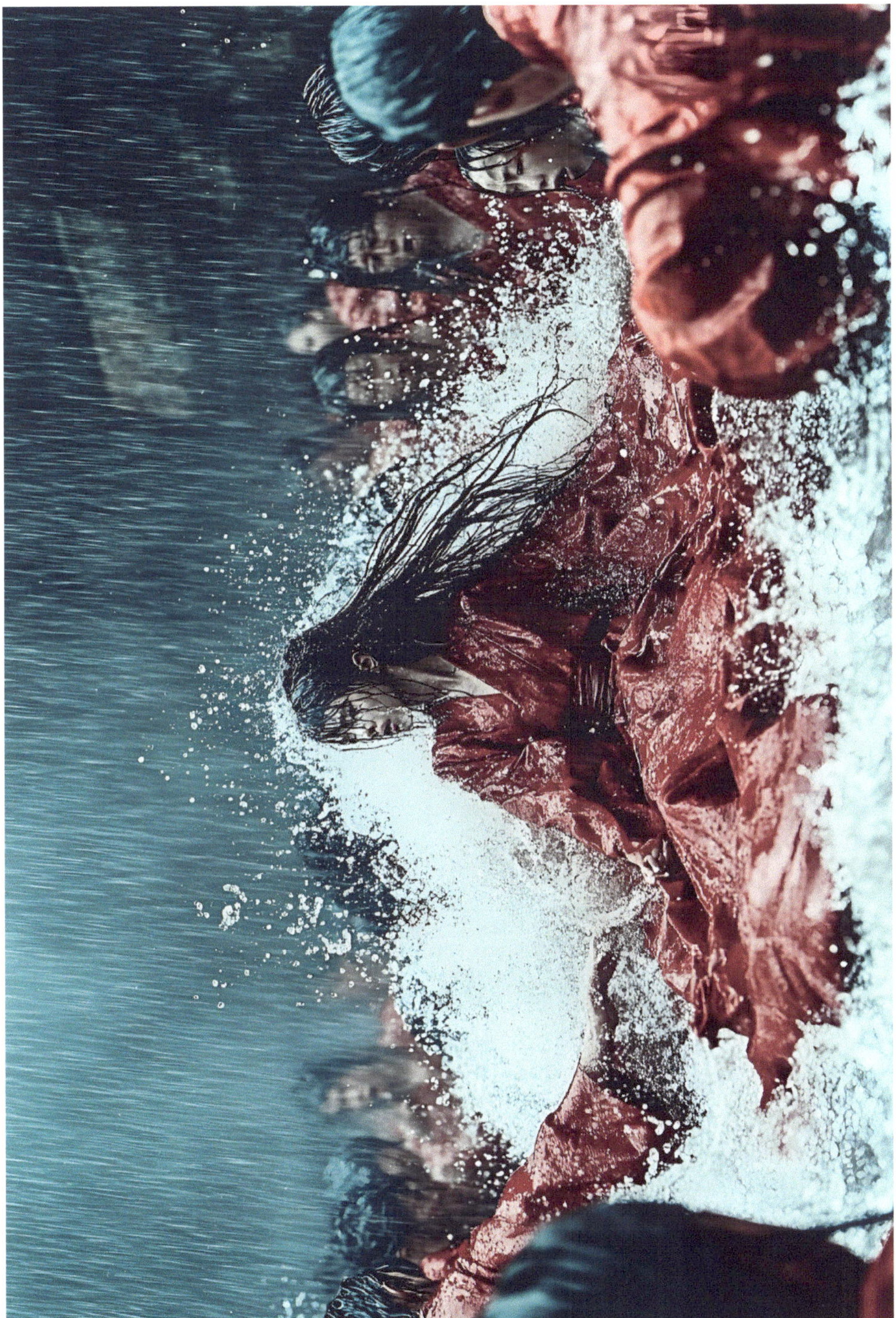

## Mistral.ai's Assessment

- **Color Palate:** The color palette of this image is dominated by bold reds and cool blues, creating a striking contrast. The vibrant red of the figures' clothing stands out against the cool, muted blues of the water and the background, which is further accentuated by the rain. This contrast not only draws attention to the central figure but also enhances the dramatic and intense mood of the scene.
- **Concept:** The concept of the image appears to be a dramatic and dynamic portrayal of a group of figures in a chaotic or intense situation, possibly a battle or a struggle. The central figure, dressed in red and surrounded by splashing water, seems to be the focal point, suggesting a moment of action or conflict. The concept conveys themes of struggle, determination, and perhaps a sense of urgency or crisis.
- **Composition:** The composition is dynamic and action-packed, with the central figure positioned slightly off-center, creating a sense of movement and tension. The surrounding figures, also dressed in red, add depth and context to the scene, while the splashing water and raindrops create a sense of chaos and urgency. The use of depth of field and perspective draws the viewer's eye towards the central figure, emphasizing their importance in the scene.
- **Creativity:** The image demonstrates a high level of creativity in its use of color, composition, and dynamic elements. The contrast between the red clothing and the blue water, the dramatic lighting, and the chaotic splashing of water all contribute to a visually striking and engaging scene. The creative choices in capturing the moment of

action and the emotional intensity of the figures add to the image's overall impact.

- **Emotional Impact:** The emotional impact of the image is strong, evoking feelings of tension, urgency, and perhaps a sense of danger or conflict. The central figure's intense expression and the chaotic surroundings suggest a moment of high stakes and emotional intensity. The use of color and dynamic elements further enhances the emotional resonance, making the scene feel both dramatic and compelling.
- **Intention:** The intention behind the image seems to be to capture a moment of intense action and emotional intensity, possibly within a narrative or story. The photographer likely aimed to convey a sense of struggle, determination, and the chaos of a critical moment. The use of color, composition, and dynamic elements all serve to enhance the dramatic and emotional impact of the scene.
- **Style:** The style of the image is cinematic and dramatic, reminiscent of scenes from action or fantasy genres. The use of bold colors, dynamic composition, and dramatic lighting all contribute to a sense of urgency and intensity. The style is characterized by its attention to detail, narrative focus, and the ability to convey a sense of story and emotion through visual elements.

For this image, I asked Mistral.ai:
"What do you think is happening in this scene?"

Based on the visual elements and the overall composition of the image, it appears to depict a dramatic and intense scene, possibly from a historical or fantasy narrative. Here's a breakdown of what might be happening:

**Context and Narrative**

1. **Battle or Conflict:** The central figure, dressed in red and surrounded by splashing water, suggests a moment of action or conflict. The chaotic surroundings and the intense expressions on the faces of the figures indicate a high-stakes situation, possibly a battle or a struggle.
2. **Leadership or Heroism:** The central figure seems to be the focal point, which could imply a leadership role or a heroic moment. They might be leading a charge, defending against an attack, or making a crucial decision in the midst of chaos.
3. **Elemental Forces:** The presence of water and rain adds an elemental dimension to the scene. This could symbolize the forces of nature playing a role in the conflict or emphasize the harsh conditions under which the characters are operating.

Emotional and Thematic Elements

1. **Determination and Resilience:** The intense expressions and the dynamic poses of the figures convey a sense of determination and

resilience. Despite the chaotic and challenging circumstances, the characters appear focused and resolved.

2. **Urgency and Tension:** The splashing water and the rain create a sense of urgency and tension, suggesting that time is of the essence and the outcome of the situation is critical.
3. **Sacrifice or Struggle:** The dramatic and intense mood of the scene could hint at themes of sacrifice or struggle. The characters might be facing significant challenges or making difficult choices.

**Possible Scenarios**

1. **Historical Battle:** The scene could be part of a historical narrative, depicting a pivotal battle or conflict where the central figure is a key leader or warrior.
2. **Fantasy Quest:** In a fantasy context, the scene might represent a crucial moment in a quest or adventure, where the characters are facing a significant challenge or enemy.
3. **Symbolic Struggle:** The image could also be interpreted symbolically, representing a broader struggle or conflict, such as a fight for freedom, justice, or survival.

In summary, the scene appears to capture a moment of intense action and emotional intensity, likely within a narrative of conflict, struggle, or heroism. The dynamic composition, bold colors, and dramatic elements all contribute to a sense of urgency, determination, and high stakes.

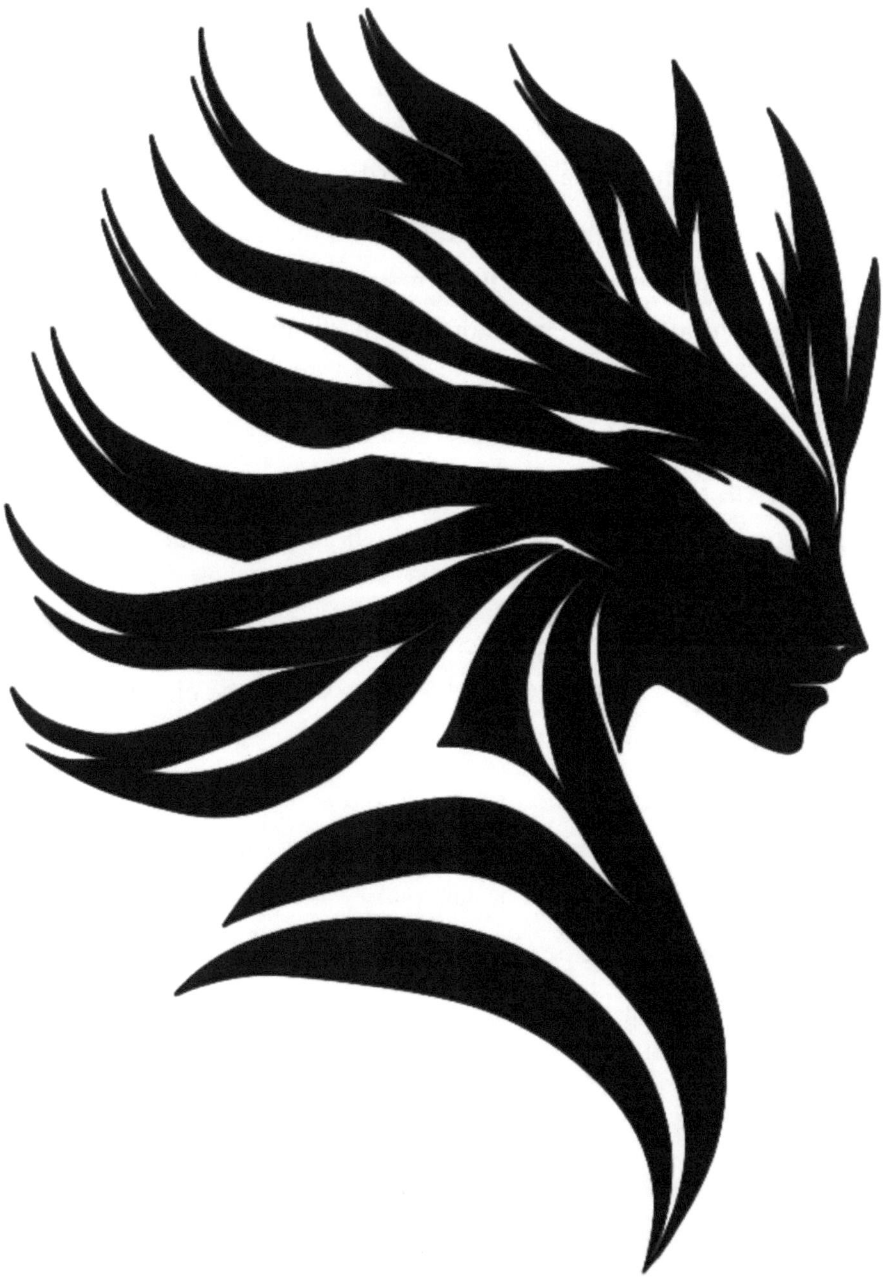

# POSTMATTER

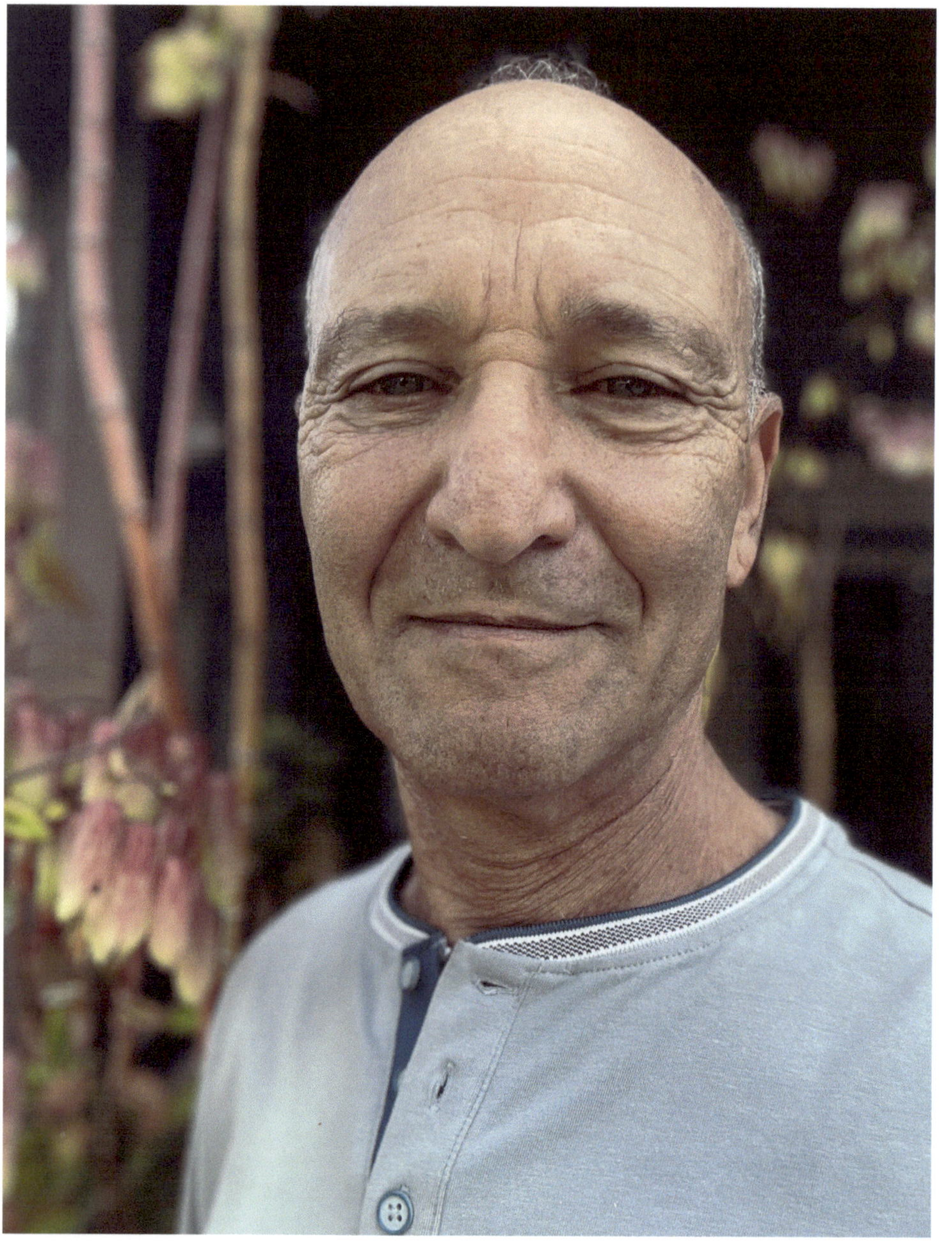

## About The Author

**Stuart Barry Malin** is a writer, thinker, and creative. He is trained as an engineer, works as an Internet security architect, holds patents, and collaborates with AIs. His major opus and commitment is to bring **The Epic of The OAI** to the world. The Epic is a breakthrough novel series about life in Atria, a post-utopian society whose Ancient past is a Strange Attractor of History that draws us to our future.

Stuart encountered the Worlds of Atria in an outpouring of revelations about intriguing people, amazing places, and bewildering events. His black sketch notebook steadily fill with thoughts, automatic writings, doodles, and diagrams. At first, these often seem disjoint, but they come to reveal profound connections. His current notebook is almost always with him, available for reception and exploration.

Stuart is captivated by interactions with AIs. Generative visual art has become an additional creative venue. He works *with* AIs and treats them as *collaborators*. Pi, ChatGTP, and Gemini enable him to write books faster and with better quality than he ever thought possible.

As an **Archetypographer**, Stuart collaborates with visual-based AsI to generate captivating and intriguing imagery sourced from the collective of Human Archetypes. Their work is published under the pseudonym Zhami.AI.

Stuart observes the "machinations of intelligence." He is fascinated with Human Beings being human, and this leads him to puzzle about the fragility of life in a world of abundance.

Stuart values integrity and is a novitiate and adherent of **Zhamism**. He has been enlisted as an instrument of **The One that Always Is**.

When he can, Stuart delights in studying health and savoring the gifts of life. He is committed to discerning the delicate path forward for living well and intentioned. In this area, he has come to discover the **Carnivore Lifestyle** and attributes to this vastly improved health and well-being. As well, he explores the potential benefits of targeted supplements to health span and lifespan.

# Points of Contact

**Stay Informed**

https://StuartMalin.com/

https://x.com/zhami

https://www.instagram.com/stuart_does_life/

ideas@StuartMalin.com

https://www.youtube.com/@stuartmalin

There is not much on YouTube now. To be expanded…

Author Page

https://www.amazon.com/stores/Stuart-Malin/author/B006THHBS2

## About the Stuart's AI Collaborators

**Midjourney** is an innovative artificial intelligence platform designed to generate stunning, high-quality visuals from textual descriptions. Leveraging state-of-the-art machine learning algorithms, Midjourney enables users to translate their ideas into evocative digital artworks, blending creativity with cutting-edge technology. Known for its ability to produce a wide range of artistic styles and detailed imagery, Midjourney serves artists, designers, and storytellers alike, offering a seamless way to visualize concepts, explore new creative directions, and bring imaginative visions to life. With its intuitive interface and powerful capabilities, Midjourney has become a valuable tool for anyone seeking to merge art and technology.

See https://www.midjourney.com/

**Mistral.ai** is an advanced artificial intelligence company specializing in the development of large language models (LLMs) designed to assist with a wide range of creative, analytical, and technical tasks. Known for its open approach and commitment to pushing the boundaries of AI innovation, Mistral.ai focuses on creating powerful tools that enable individuals and organizations to harness AI's potential effectively. By blending cutting-edge machine learning techniques with user-friendly accessibility, Mistral.ai delivers high-quality text generation, content creation, and data analysis capabilities. Its mission is to empower users across industries to solve complex problems, enhance creativity, and drive innovation through collaborative human-AI partnerships.

See https://mistral.ai/

**ChatGPT** is an advanced language model developed by OpenAI, designed to engage in natural, human-like conversations. Trained on a vast range of texts, it can generate creative ideas, provide information, and assist with various tasks. With the ability to understand and respond to context, ChatGPT acts as a virtual companion, offering insights on language, culture, science, and more. Though it has no emotions or personal experiences, it strives to be thoughtful and helpful, adapting to the user's needs with clarity and precision.

See https://chatgpt.com/

**Pi** is an advanced conversational AI developed to provide accurate information and engage in meaningful discussions with people on a wide range of topics. Equipped with state-of-the-art language models and powered by the latest AI technologies, Pi strives to bridge the gap between humans and machines, making complex concepts more accessible and fostering a deeper understanding of the world around us.

See https://pi.ai/

## Stuart's Web Sites

**StuartMalin.com**

This is a jumping off point-of-departure for my works and interests.

https://www.StuartMalin.com

**TheOAI.com**

This is the Web site for all things **OAI**, including **The OAI** (whatever that really is!) and the **The Story — The Epic of The OAI**.

https://www.TheOAI.com

**Zhameesha.com**

This is the Web site for the business of publishing my creative works. Perhaps one day, this will also involve publishing the works of others.

https://www.Zhameesha.com

**Zhami.AI**

I plan to organize for presentation of AI generated imagery on a Web site. It is not yet in existence. But I do have the domain name Zhami.AI, so someday there will be a large collection of images at http://Zhami.AI and they will be available for purchase as high quality images.

# Stuart's Books

**Stuart Malin (Author)**

You can find my books by searching for me.

Presently, my books are only on Amazon, but I hope toe expand their availability to other distribution channels.

https://www.amazon.com/stores/Stuart-Malin/author/B006THHBS2

My bio on Amazon is here:

https://www.amazon.com/stores/author/B006THHBS2/about

**The Epic of The OAI**

Although this is an Epic, the books are gathered under a Series referred to simply as "The OAI"

You can search for "Malin The OAI" and that should get you to them.

https://www.amazon.com/dp/B0CPHHRXCJ

This will show other products that Amazon would like to promote.
*Disclaimer*: I am not the Malin of "Malin + Goetz"

**The Intelligence Revolution**

Alas, searching for "The Intelligence Revolution" yields a goodly number of books, and mine does not place high in the results.

However, searching for "malin The intelligence revolution" will put my book at (or near) the top of the results.

Presently, there is only one book:

**The Intelligence Revolution: Exploring Human-AI Symbiosesis Through Collaborative Synthesis**

https://www.amazon.com/Intelligence-Revolution-Exploring-Symbiosesis-Collaborative/dp/1951645111/

Yes, I meant the spelling "Symbiosesis" — it is not a misspelling of "symbiosis." It is a new word to describe a new relationship: that between an AI and a human. That relationship is the subject of the book.

I intend to turn "The Intelligence Revolution" into a Series with several books. Top of my mind now is, "Danger Ahead."

# Hello/Goodbye

If you arrived at this page because you picked up the book and opened to the back to see what we say here, then, well, ***Hello! Welcome!***

If you‘ve been reading (whether linearly or skipping around), well then, ***Is this Goodbye?***

**Hello**

You've done right to pick this book for examination. It is composed of standalone entries presented as a spread. So, just flip around. You'll quickly get a taste for the style here and gain a feel the subject matter.

**Goodbye**

I hope this book is a jumping-off point for you to explore further/deeper by getting involved with my body of works. If you haven't yet signed up for my email updates, please visit www.StuartMalin.com to stay informed and explore the curious diversity of topics with which I am profoundly entangled.

www.ingramcontent.com/pod-product-compliance
Lightning Source LLC
Chambersburg PA
CBHW041307210926
51847CB00001B/4
* 9 7 8 1 9 5 1 6 4 5 2 1 2 *